A BRAIN FOR INNOVATION

A Brain for Innovation

THE NEUROSCIENCE OF IMAGINATION AND ABSTRACT THINKING

Min W. Jung

Columbia University Press
New York

Columbia University Press
Publishers Since 1893
New York   Chichester, West Sussex
cup.columbia.edu

Library of Congress Cataloging-in-Publication Data
Names: Jung, Min W., author.
Title: A brain for innovation : the neuroscience of imagination and
abstract thinking / Min W. Jung.
Description: New York : Columbia University Press, [2024] |
Includes bibliographical references.
Identifiers: LCCN 2023018976 (print) | LCCN 2023018977 (ebook) |
ISBN 9780231213363 (hardback) | ISBN 9780231559850 (ebook)
Subjects: LCSH: Cognitive neuroscience. | Imagination—Physiological
aspects. | Abstraction—Physiological aspects. | Brain—Physiology.
Classification: LCC QP360.5 .J86 2024 (print) | LCC QP360.5 (ebook) |
DDC 612.8/233—dc23/eng/20230822
LC record available at https://lccn.loc.gov/2023018976
LC ebook record available at https://lccn.loc.gov/2023018977

Cover design: Henry Sene Yee
Cover image: Shutterstock

CONTENTS

ACKNOWLEDGMENTS

I'm grateful that I've had the opportunity to work with so many great fellow scientists over the course of my career, including my doctoral mentor, Gary Lynch, and postdoctoral mentor, Bruce McNaughton, who have shaped the way I view how the brain supports high-level mental functions. I want to thank all of these scientists, but there are too many to mention by name here.

A special note of appreciation is owed to my former students Jong Won Lee, Hyunjeong Lee, Sung-Hyun Lee, and Youngseok Jeong, as well as my collaborators Daeyeol Lee, Inah Lee, Woong Sun, and Woonryung Kim, who made significant contributions to the works discussed in chapters 5 and 6 and the appendixes. Special thanks also go to Kyoon Huh and Eunjoon Kim as colleagues for their advice and support.

I want to thank Se-Bum Paik for his insightful comments on the first draft of chapter 11 ("Deep Neural Network"). I'd also like to thank Maame Boetamaa for her editorial help, Jong Won Lee for preparing figures, Jooyong Shin for assisting with references, and Verner Bingman and Kimberly Wade for providing original images.

The staff at Columbia University Press has been extremely helpful to me in finishing this book. In particular, the project was greatly fueled by the enthusiasm of Miranda Martin, the project's editor. She and the rest of the production, graphics, and marketing team at Columbia University Press have my appreciation.

Most of all, I would like to express my deepest gratitude to my parents and my family, Inhee, Alice, and Amy, for their sacrifice, support, and encouragement throughout my life as a scientist.

A BRAIN FOR INNOVATION

INTRODUCTION

Humans are extraordinary animals. As the only known species that can understand the concept of their own existence, humans have long pondered their place in the world and their relationship to the universe. These inquiries have resulted in some of the most significant advances in science, philosophy, and spirituality. Our extraordinary capacity for introspection and self-reflection distinguishes us from all other living species on the planet. Despite our exceptionality, we share with other living organisms the same ultimate goal: the survival and perpetuation of our species.

Which animals best achieve the ultimate biological goal? Without a doubt, insects; they are currently the most adapted animals on earth. Scientists estimate their number to be about ten quintillion (10,000,000,000,000,000,000), and together they weigh seventy times more than all humans combined. Their diversity is also amazing. There are about 1 million known insect species, and this amounts to around 90 percent of all known animal species and more than 50 percent of all known life species.[1] Moreover, the number of unknown insect species is estimated to be between 2 and 10 million (5.5 million on average).[2] It is not surprising that there exists a discipline, entomology, dedicated to studying insects: insects are the most dominant animals on earth in terms of number, mass, and diversity. Scientists predict that insects will survive tenaciously when many other animal species, including humans, become extinct in the distant future. Biologically, it is obvious that humans are not the most successful animal species on earth.

Even though we are not currently the most successful animals on earth, we are perhaps the most successful *large* vertebrate animals. We have managed to thrive in a wide range of environments on every continent, and our total weight now far exceeds that of all wild land vertebrate animals. More importantly, we are the only animals that have advanced technologies, created sophisticated cultures, established large societies, and, remarkably, gained power to affect and shape our environment in ways that no other animal has.

These achievements would not have been possible without our capacity for innovation. We tend to try new things to improve our futures. The accumulation of innovations big and small throughout history has eventually enabled us to build civilizations on a global scale. We even have established a social system to promote innovations—the patent laws. In this respect, humans may be regarded as innovative animals or *Homo innovaticus*.

Figure 0.1 shows advances in technology along with the growth of the world's population since the dawn of agriculture. The population size grew exponentially, and technological changes accelerated at an astonishing rate during the last two centuries. The major technological events shown represent only a minute fraction of the scientific and technological advances mankind has achieved during the last two hundred years.

Why are we so innovative? Perhaps our brains differ from those of other animals. But how? This is the main topic of this book. Innovation requires a new insight, and a critical factor for new insights is the capability for imagination. We come up with new technologies, bits of knowledge, ideas, and art by combining existing ones in new ways using our imagination. Albert Einstein referred to the power of imagination this way: "Logic will get you from A to B. Imagination will take you everywhere." A question arises then as to whether the capability for imagination is unique to humans. The answer to this question is clear: "Definitely not." Imagination is far from being a unique human mental faculty. Psychological, neuroscientific, and animal behavioral studies have provided converging evidence that animals are capable of imagination. In particular, neuroscientific research in the last two decades has revealed neural activity seemingly related to imagination.

What then is the unique human mental faculty that enabled innovation throughout human history? Imagination promotes but does not guarantee innovation. The scope of imagination is constrained by one's cognitive capacity. Without a sufficiently high cognitive capacity, the content of one's imagination would be far from being innovative. In particular, the capacity

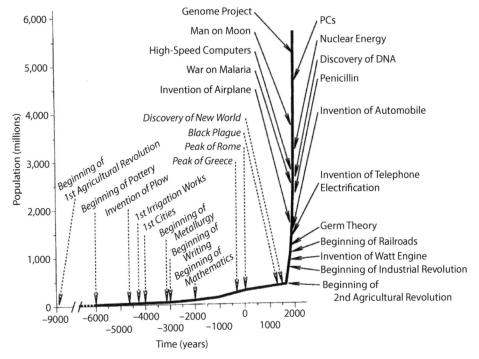

FIGURE 0.1. Population growth and major technological events. Figure reproduced with permission from Robert W. Fogel, "Catching up with the Economy," *American Economic Review* 89, no. 1 (1999): 2 (copyright American Economic Association).

for high-level abstraction is essential for innovations involving conceptual knowledge such as those shown in figure 0.1. In other words, humans are particularly innovative because they have the unique capacity to imagine freely using high-level abstract concepts such as imaginary numbers, vectors, gravity, atoms, genes, mutations, algorithms, beauty, humanism, free will, liberty, and social justice.

I by no means argue that abstract thinking is a unique human mental faculty. It is well known that other animals, especially primates, are capable of abstract thinking. However, no animals come close to humans in terms of the level of abstraction. Humans are superb at forming and manipulating high-level abstract concepts and imagining freely using them. A prime example of this capability is language. Only humans possess a true capability for language with specialized brain regions dedicated to language processing. Innovations such as those shown in figure 0.1 would have not been possible without the unique human capacity for imagination using high-level abstract concepts. Thus, of a diverse array of advanced mental faculties, one critical element for innovation would be the capacity to imagine freely using high-level abstract concepts.

Our current understanding of the brain mechanisms that allow this is limited. Nevertheless, we have a sufficiently large body of discoveries from several different disciplines, neuroscience in particular, to allow discussion of the specific neural processes underlying this great human faculty. Neuroscience has traditionally focused on how the brain processes and stores external sensory information and controls behavior using this information. In contrast, the neural processes underlying internal thinking and self-generated thought, especially those related to imagination and creativity, have received less attention. This trend is changing because of discoveries such as the existence of a neural system that is particularly active when we are engaged in internal thinking such as daydreaming and envisioning futures; that this neural system interacts dynamically with other neural systems during creative thinking; that the hippocampus, already known to play a crucial role in encoding new memories, also plays an important role in imagination; and neuronal activity identified in animal studies that is seemingly related to imagining future episodes. These discoveries allow us to glimpse into the neural processes that encompass internal mentation, imagination, creative thinking, and innovation.

This book will delve into modern neuroscientific research by examining various discoveries that have provided important insights into the neural mechanisms underlying imagination and high-level abstraction. I intend to take this matter down to the level of neural circuit operation wherever possible rather than merely assuming some brain region or homunculus is doing the job. Groundbreaking discoveries in the last two decades centered around the hippocampus have enabled us to explain the process of imagining the future at the level of neural circuits. Even though we have less understanding of the neural basis of high-level abstraction, some clues allow us to conjecture about the neural network processes that underpin high-level abstraction. Additionally, I will highlight related works in psychology, anthropology, and artificial intelligence. In doing so, I will try to explain a unique faculty of *Homo sapiens*, the capacity for innovation, in terms of the organization and functioning of neural systems and circuits.

This book has four sections. The role of the hippocampus in imagination is covered in part 1 (chapters 1–3). The hippocampal neural circuit processes underlying imagination are examined in part 2 (chapters 4–7). Part 3 (chapters 8–11) moves to the neural basis of high-level abstraction in humans. Part 4 (chapters 12–14) takes us beyond imagination to help us understand creativity and how humans might use the capacity for innovation in the future.

Note that many significant discoveries that are pertinent to the book's numerous sections have been omitted. This book aims to bring together disparate findings in neuroscience and related fields in order to explain the neural basis of innovation in a concise manner. Inevitably, topics and findings are handled in a highly selective manner. There is a huge amount of scientific literature available, and many sources are beyond the scope of this book. Also note that this book's writing style differs substantially from that of scientific monographs. Even though I delve deep into modern neuroscientific research, I intend to present a concise narrative to an intelligent reader who has no expert knowledge of neuroscience.

I hope that this book will help you gain a better understanding of the neural processes that underlie one of the most fascinating and essential aspects of human nature: our capacity for innovation.

PART I

Hippocampus and Imagination

Chapter One

HIPPOCAMPUS

From Memory to Imagination

What is the neural basis of the ability for unbounded imagination using high-level abstract concepts? Surprisingly, the neuroscientific journey to find an answer to this question begins with the study of memory. Remembering past experiences is one thing, and imagining future events is another. Therefore, one would presume that the neural machinery for imagination differs from that for memory. In fact, that's what neuroscientists used to think until 2007, when scientists found an overlap in the brain regions in charge of memory and imagination. In particular, the hippocampus, which is well known for playing a critical role in encoding new memories, was found to be involved in imagination as well. In this chapter, we examine landmark discoveries on the hippocampus beginning from its role in memory (1950s) to its role in imagination (2000s).

HENRY MOLAISON: AN UNFORGETTABLE AMNESIAC

Henry Molaison, known by his initials, H.M., to the public until his death, was born in February 1926 in Manchester, Connecticut. He suffered from such severe epilepsy that he could not lead a normal life by the age of twenty-seven. In September 1953, William Scoville, a neurosurgeon, removed parts of Molaison's brain to alleviate his symptoms. The surgery was effective in controlling the seizures. However, an unexpected side effect of the surgery deprived him of a vital brain function—remembering new experiences.

Surprisingly, other functions, such as sensation, movement, language, intelligence, short-term memories, and even old memories, were barely compromised. It appeared that only the ability to remember new experiences was profoundly impaired.

Molaison's case indicates that a separate neural system is in charge of encoding new memories. Before this case, many scientists thought that memory was a function of the entire brain rather than a specific brain structure. Consider the well-known work of Karl Lashley. After training rats to run in a maze, Lashley made various cuts on their brains to impair their memories. However, his experiment failed to find the specific brain region that, when cut, impaired the rat's memory as assessed by their behavior of running to the goal box. Instead, he found that the degree of memory impairment correlated with the degree of knife cuts. He proposed then, based on these observations, that the whole brain has the capacity to store memory.[1]

In contrast to this proposal, the Molaison case clearly demonstrates that a separate brain region specifically oversees encoding new memories without being involved in many other brain functions. Additionally, Molaison's older memories' remaining intact indicates that separate neural systems are in charge of encoding new memories and storing long-lasting memories. During his surgery, his medial temporal lobe, including the hippocampus, was removed bilaterally (see fig. 1.1). The results of this case indicate that while the medial temporal lobe is in charge of encoding new memories, it is not the final memory storage site. These astonishing discoveries had deep impacts on our understanding of memory and the brain. Molaison lost his memory but left an unforgettable legacy in neuroscience.

MEMORY CONSOLIDATION

Another unexpected finding from Molaison's case was his development of temporally graded retrograde amnesia. He lost not only the ability to form new memories (referred to as *anterograde amnesia*) but also the ability to recollect some of the memories for events he experienced before the surgery (referred to as *retrograde amnesia*). His retrograde amnesia was temporally graded; his recent memories were more likely to be lost while distant memories were spared. In fact, he could not remember most of the events he experienced a year or two before the surgery. This indicates that the medial temporal lobe is necessary not only to encode new memories but

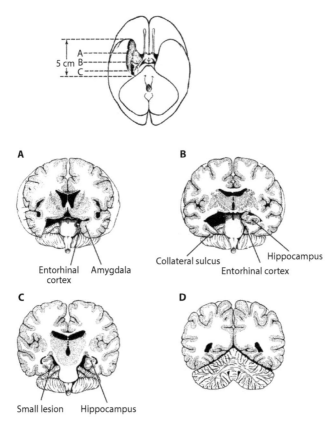

FIGURE 1.1. Drawings of Henry Molaison's brain. The medial temporal lobe was removed bilaterally, but the right hemisphere is left intact here to show removed structures. Figure reproduced with permission from Suzanne Corkin et al., "H. M.'s Medial Temporal Lobe Lesion: Findings from Magnetic Resonance Imaging," *Journal of Neuroscience* 17, no. 10 (May 1997): 3965 (copyright Society for Neuroscience).

also to recollect memories of recently experienced events. According to the *systems consolidation theory*, which is currently the most popular theory on brainwide organization of memory, new memory is rapidly stored in the hippocampus and then goes through a "consolidation" process so that it is eventually stored elsewhere in the brain, such as the neocortex.[2]

Graded retrograde amnesia and memory consolidation indicate that the way humans encode and store experiences as memories is different from that of a digital computer. Why then do we store memory in this way? Why not simply send memories to the final storage site? Probably because it is advantageous, albeit cumbersome, to have two separate memory storage sites. On the one hand, it would be useful to remember details of experienced events to make better choices in everyday life. On the other hand, we may run out of storage space if we store most of our experiences as permanent memories. One way to resolve this conundrum would be having two memory storage sites: one for the temporary storage of details of experiences and another for the permanent storage of the gist of experiences.

Suppose you commute to work by car. You need to remember exactly where you parked your car in the morning to get back to it in the evening. However, it wouldn't be useful to remember your exact daily parking locations for the rest of your life. Instead, remembering that you drove to work and probably parked your car in the company's parking lot would be less wasteful as a long-lasting memory. The hippocampus may temporarily store detailed memories of recent experiences (this type of memory is called *episodic memory*) whereas the neocortex may gradually extract general facts from ensembles of experiences and store them as permanent memories (*semantic memory*).[3]

Systems consolidation is not the only existing theory on the organization of memory. According to the *multiple trace theory*, the hippocampus stores remote memories even after consolidation;[4] the hippocampus is not necessary to recollect memories of general facts (semantic memories) but is necessary to recollect memories of specific experienced events (episodic memories) even after consolidation.[5] As such, there are multiple theories on how memories are organized in the brain, indicating that we do not yet perfectly understand why and how initially acquired memories are consolidated over time to become permanent memories. In fact, memory consolidation is directly related to the central issue of this book, imagination. I think that memory consolidation is a process of actively selecting and reinforcing valuable options for the future by recombining past experiences using imagination rather than passively storing incidental events. I also think that this is a fundamental basis for the human capacity for innovation. I will discuss this matter in more detail in chapter 6.

Another important finding from Henry Molaison's case is that there are multiple forms of memory. Molaison could not remember new facts and events but could learn to perform new tasks by practice. This indicates that remembering new facts and events (*declarative* or *explicit memory*) is mediated by the medial temporal lobe, whereas learning new skills such as riding a bicycle (*procedural* or *implicit memory*) is mediated by other brain structures. Despite a large body of studies on multiple forms of memory, this book does not further explore these memory forms as they do not directly relate to the main issue at hand.

HIPPOCAMPUS AND IMAGINATION

The Molaison case, first published in 1957, changed the course of memory research.[6] Fifty years later, in 2007, new findings were published that once again changed the course of memory research. In one study, Demis Hassabis and colleagues demonstrated that hippocampal amnesiacs have trouble not only in memory encoding but also in vivid imagination.[7] They asked hippocampal amnesic patients and a normal control group to imagine plausible events under hypothetical situations. Some sample verbal cues for imagination are "Imagine you are lying on a white sandy beach in a beautiful tropical bay" and "Imagine that you are standing in the main hall of a museum containing many exhibits."

The results of this experiment were surprising. Hippocampal amnesic patients had trouble constructing new imagined experiences. The control subjects, of course, came up with diverse imaginary scenarios with little difficulty. Try this exercise yourself. You can probably easily imagine a plausible sequence of events without much effort. However, hippocampal amnesic patients have trouble vividly imagining plausible episodes. In other words, damage to the brain structure known to play a critical role in memory can also impair the ability for vivid imagination.

Two other studies published in 2007 yielded the same conclusion using a different approach.[8] They used functional magnetic resonance imaging (fMRI), a widely used brain imaging technique, to examine which brain areas are activated during imagination in neurotypical people. The hippocampus was activated when the subjects were recollecting autobiographical memories, which is expected; the hippocampus is crucial for remembering recent experiences. Surprisingly, the hippocampus was also activated when the subjects were envisioning the future. In other words, the hippocampus was active not only during remembering past episodes but also during imagining future events.

DEFAULT MODE NETWORK

To better understand hippocampal activation during imagination in a broader context, consider that certain brain areas are particularly active when we are relaxed and not paying much attention to the external world. These areas are collectively called the *default mode network*. Scientists have traditionally focused on how the brain processes information in response to external sensory stimuli, but little attention has been paid to brain activity during its idle state. Surprisingly, scientists have found that certain brain regions are more active during rest than while performing attention-demanding tasks.[9]

The discovery of the default mode network was, in fact, accidental. Brain imaging techniques began to be widely used to investigate brain functions in the 1980s, and early studies commonly employed passive conditions, such as rest, as a control for task-performance conditions. As experimental data accumulated, scientists realized that there are certain brain areas that are more active during passive conditions. The name Marcus Raichle and colleagues collectively gave these brain regions was because they represent baseline-state (default) brain activity.[10]

What were the subjects doing during the passive conditions? They reported that they were recollecting autobiographical memories (e.g., thinking about the previous night's dinner) or envisioning the future (i.e., daydreaming) during rest. Subsequent studies have found that the default mode network is active in association with diverse mental activities such as thinking about someone else's thoughts (called *theory of mind*), making moral decisions (e.g., the trolley dilemma shown in figure 1.2), and performing creative thinking tasks (e.g., the divergent thinking task in which subjects are asked to produce multiple alternative options in response to such an open-ended question: "In what ways can a brick be used?").[11] These results suggest that the default mode network is activated in association with internal mentation. The brain appears to be equipped with a task-associated network, which is active when we are engaged in an activity that requires us to pay attention to the external world, while the default mode network is active when we are engaged in internal mentation.

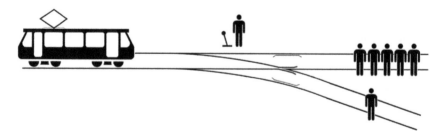

FIGURE 1.2. Trolley dilemma. A runaway trolley is hurtling toward five workers who are unaware of the trolley coming. You can save their lives by pulling a lever to divert the trolley to a sidetrack. However, there is a lone worker in the sidetrack who is also unaware of the trolley coming. What would you do? Would you kill one person to save five? Figure reproduced with permission from McGeddon, "File:Trolley Problem.svg," Wikimedia Commons, updated March 6, 2018, https://commons.wikimedia.org/wiki/File:Trolley_Problem.svg (CC BY-SA 4.0).

As outlined in the rest of this book, innovative ideas often emerge in resting states and even during sleep. The brain may seem to be idle when we take a rest. On the contrary, the default mode network is active while we rest, and innovative ideas may appear while our minds wander freely. We will examine this issue closely in chapter 13. For now, germane to the rest of these discussions is noting that the hippocampus is a main component of the default mode network.

In summary, hippocampal damages impair one's ability to imagine, and the hippocampus, as an important part of the default mode network, is activated in association with internal mentation including envisioning the future. These findings are not without caveats, however. Patients with hippocampal damages usually have damage in other brain areas as well. Also, brain imaging studies rely on the indirect measure of brain blood flow instead of neural activity. Nevertheless, the implication of these findings is clear; the hippocampus is involved not only in memory but also in imagination. We reached a new turning point in memory research fifty years after the publication of the Henry Molaison case. For this reason, the journal *Science* named the discovery of the hippocampus's involvement in imagination as one of ten breakthroughs in 2007.[12]

Chapter Two

FALSE MEMORY

It is now clear that the same brain structure, the hippocampus, is involved in memory as well as imagination. A question then arises as to whether the imagination function of the hippocampus may interfere with its memory function. Wouldn't it be difficult to maintain memory's integrity if it were influenced by imagination? Isn't there a danger of mixing up what we imagined with what we have experienced? In fact, it is well known in psychology that memory is not static but changes over time, and we may even remember events that never happened. The phenomenon of *false memory* (a recollection that is partially or fully fabricated even though it seems real) raises the possibility that brain regions related to memory and imagination may closely interact with each other. Now, the finding that the hippocampus is involved in both memory and imagination provides a potential account for the neural mechanisms underlying false memory.

WHAT I REMEMBER VERSUS WHAT
ACTUALLY HAPPENED

We do not recollect events exactly as we experienced them. Most often, our recollection of an event differs substantially from the actual event. This is because the way the brain processes and stores information differs from how a computer works. For example, we tend to extract the gist and meaning from our experiences so that what we infer may be mixed up with what we actually experience. If we hear a list of words related to sleep (e.g., bed, rest, awake, etc.), we tend to recollect "sleep" as being on the list, even though it is not.[1] Also, unlike a computer recovering a file, some memories can interfere with the retrieval of other memories.[2] If a long time has elapsed since experiencing a particular event, and if you have experienced similar events several times since the original experience, chances are you would find it difficult to recollect the original event exactly as it happened; your memory of the original event may get mixed up with memories of related events.

The brain also tends to fill in missing information. Let's pretend you were involved in a traffic accident. You will likely think about the accident from time to time because it is an extraordinary event. During this process, your brain may fill in missing information. For example, even though you initially do not recollect the color of the other driver's clothes because you did not pay attention to it, your brain may fill in this information as you recall the episode afterward. The filled-in information will be strengthened as you repeat this process. As a result, you may vividly recollect that the other driver was wearing a red shirt even though he was in fact wearing an orange shirt. Why red? You may have met many people with red shirts recently. Or you may have watched a movie in which the villain was wearing a red shirt while driving a car. As in this hypothetical example, we tend to fill in missing information with our best guesses based on our expectations and past experiences, which explains why eyewitness testimony can be dangerously unreliable.[3]

THE GEORGE FRANKLIN CASE

Occasionally, we may form a memory of an event that never happened at all. Here, we will examine two suspected cases of such false memory that captured public attention. The first is the case of George Franklin, which originated from a murder in Foster City, California, in 1969.[4] An eight-year-old girl, Susan Nason, was raped and murdered, and the case remained unsolved for years. Twenty years later, Eileen Franklin, a friend of Susan Nason, accused someone of the crime. Surprisingly, the accused was her father, George Franklin. Eileen testified in court that she witnessed George crushing Susan's skull with a heavy rock. She said that a vivid memory of the horrible scene, which had been forgotten for twenty years, suddenly flashed into her mind as she was playing with her young daughter, Jessica.

Eileen Franklin's testimony initiated a controversy between *repressed* memory and *false* memory. Repressed memory is a concept originally postulated by Sigmund Freud as a defense mechanism against traumatic events. The mind may unconsciously block access to horrific memories, which may be recovered years later. This idea was widespread among clinical psychologists and ordinary people at the time of George Franklin's trial. With no physical evidence other than his daughter's testimony, the jury found him guilty of first-degree murder, and he was sentenced to life imprisonment.

Contrary to the jury's verdict, the following discussion argues for false memory rather than repressed memory. First, the postulated repressed memory has not been proven scientifically. Second, memories change over time. It is therefore questionable whether Eileen could recollect details of an event exactly as it happened twenty years earlier. Third, all the details she provided about the murder could be found in newspaper articles. Furthermore, some details were incorrect in both the newspaper articles and Eileen's statement. For example, one article mistakenly stated that the ring Susan was wearing on her right hand contained a stone (the ring with a topaz stone was in fact on her left hand), and Eileen made the same error. Fourth, it was revealed later that Eileen recalled the memory through hypnosis. Hypnotically refreshed testimony is not allowed in court because the witness's memory can be contaminated with misinformation. Correct and incorrect information may get mixed into a coherent narrative during hypnosis, which increases the confidence of the witness about the refreshed memory. Fifth, Eileen later accused her father of two additional cases of rape and murder (Veronica Cascio and Paula Baxter), but DNA tests cleared him of the charges. This called into question the credibility of Eileen's testimony. In 1996, he was released from prison. In 2018, another man, Rodney Lynn Halbower, received life sentences for those two murders.[5]

THE PAUL INGRAM CASE

In 1988, Paul Ingram, who was then the chief civil deputy of the Thurston County Sheriff's Office in Olympia, Washington, was accused by his two daughters, twenty-two-year-old Ericka and eighteen-year-old Julie, of sexually abusing them for years when they were children. He initially denied the charges but eventually admitted that he sexually abused his daughters during satanic rituals. Since he pleaded guilty, it seemed that the case was resolved with no controversy.

This seemingly simple case unfolded into a complicated tale, however. Ericka's accusations were gradually amplified to grandiose stories of ritual abuse that included baby sacrifices, for which the police failed to find physical evidence. For example, she insisted that she watched twenty-five babies being sacrificed during the rituals and that they were buried in the family's backyard. However, no human bones (only a fragment of an elk bone) were found on the property. In fact, months of police investigation, the most extensive in the county's history, yielded no physical evidence whatsoever for the claimed rituals or homicides.[6]

What about Paul's recollection of satanic rituals? A "little experiment" performed by Richard Ofshe, a social psychologist from the University of California, Berkeley, suggests that his confessions may be "coerced-internalized false confessions."[7] Ofshe joined the investigation as an expert in cults and mind control, but he was also interested in coercive police interrogations. He introduced a false allegation that Paul forced his son and daughter to commit incest with each other. Paul initially denied this accusation but eventually wrote a detailed confession on it. He even refused to believe that the incident was false when Ofshe revealed that he had made it up.[8] It is still unclear whether Paul's confessions reflect what happened, false memories, or a mixture of both.[9] Nevertheless, together with the George Franklin case, Paul Ingram's case contributed to the public awareness that misinformation and suggestions by authoritative figures, such as trusted psychologists or police officers, may implant false memories in the psyche of susceptible people.

LOST IN A SHOPPING MALL

The two preceding cases suggest that we may fabricate memories of events that never actually happened. Elizabeth Loftus, a psychology professor at the University of Washington, provided expert testimony on this possibility in the George Franklin trial. Despite her testimony, he was convicted in 1990. Her studies until then proved that memories could be altered by information received after an event (known as the *misinformation effect*) but did not prove that memories could be wholly invented.[10]

To prove the possibility of totally fabricated event memories, Loftus, with her student Jacqueline Pickrell, ran an experiment—the "lost in a shopping mall" study—and published the results in 1995.[11] Twenty-four subjects were given four short stories about events from their childhood and instructed to record details of their memories of them. They were then interviewed twice, with each interview conducted over a span of one to two weeks. Three of the four stories were true, but one was a false event about getting lost. For example, the following summary was given to a twenty-year-old Vietnamese American woman: "You, your mom, Tien, and Tuan all went to the Bremerton K-Mart. You must have been 5 years old at the time. Your mom gave each of you some money to get a blueberry Icee. You ran ahead to get into the line first and somehow lost your way in the store. Tien found you crying to an elderly Chinese woman. You three then went together to get an Icee."

On the one hand, most people (18 of 24 or 75 percent) denied having the memory of getting lost and successfully remembered 49 of the 72 true events (a 68 percent success rate). The amount of details, clarity rating, and confidence rating were also much greater for the memories of the true events, indicating that our event memories are reasonably reliable.

On the other hand, six subjects (25 percent) said they remembered, fully or partially, the event of getting lost in their childhood, showing that it is possible to implant, by misinformation and suggestions, false memories of events. The following is an example from one subject, who falsely believed that she had truly been lost during the second interview: "I vaguely, vague, I mean this is very vague, remember the lady helping me and Tim and my mom doing something else, but I don't remember crying. I mean I can remember a hundred times crying. . . . I just remember bits and pieces of it. I remember being with the lady. I remember going shopping. I don't think I, I don't remember the sunglass part."

A PICTURE IS WORTH A THOUSAND LIES

The result of the "lost in a shopping mall" study has been replicated in subsequent studies. Let's examine one that employed a fake visual image.[12] Twenty subjects who had never taken a hot-air balloon ride before were presented with a doctored photograph showing them riding in a hot-air balloon (see fig. 2.1). They were then interviewed three times over the next seven to sixteen days. By the third interview, ten subjects (50 percent) claimed that they remembered at least some details of the balloon ride (consistent elaboration of information not shown in the fake photo was required to be

FIGURE 2.1. Photoshopped images generated for use in a false memory study (courtesy of Kimberly Wade).

qualified as a false memory). The following is from the third interview with a subject who was rated to have developed a partial false memory:

Interviewer: Same again, tell me everything you can recall about Event 3 without leaving anything out.

Subject: Um, just trying to work out how old my sister was; trying to get the exact . . . when it happened. But I'm still pretty certain it occurred when I was in form one (6th grade) at um the local school there. . . . Um basically for $10 or something you could go up in a hot air balloon and go up about 20 odd meters . . . it would have been a Saturday and I think we went with, yeah, parents and, no it wasn't, not my grandmother . . . not certain who any of the other people are there. Um, and I'm pretty certain that mum is down on the ground taking a photo.

This study corroborates the conclusion that we may form false memories of events. More subjects developed a false memory in the hot-air balloon study than in the lost-in-the-mall study. This is perhaps because the subjects saw a clear, albeit fake, visual image (see fig. 2.1). After all, isn't a picture worth a thousand words? The authors metaphorically titled this study "A Picture Is Worth a Thousand Lies."

CONSTRUCTIVE MEMORY

We started this chapter with the issue of potential interactions between memory and imagination. Given that the hippocampus is involved in both memory and imagination, isn't it possible to mix up what we imagined with what we have actually experienced? The answer is clear. Imagination does influence memory, so it is even possible to implant fabricated memories under certain circumstances. Pretend you are one of the study's participants: It seems to be true that you once got lost in a shopping mall during the childhood because your mom and brother said so. You don't have a clear memory of that event, but you were asked to recall and write down episodic details of it. Here, imagination may slip into the recalling process, gradually filling in episodic details of a false memory.

Contrary to common sense, memory and imagination may not be two independent processes; our memory clearly relies on constructive processes that are sometimes prone to error and distortion. Daniel Schacter, a psychologist at Harvard University, named this aspect of memory *constructive memory*: "When we remember, we piece together fragments of stored information under the influence of our current knowledge, attitudes, and beliefs."[13] Imagination is also a process of piecing together fragments of stored information. If so, it would be more efficient for the brain to share a common constructive process for memory and imagination rather than maintaining two independent processes. From this perspective, it would not be surprising to learn that the hippocampus is involved in both memory and imagination. Although it is not favorable for remembering an event precisely as it happened, it is adaptive in that it "enables past information to be used flexibly in simulating alternative future scenarios without engaging in actual behaviors."[14]

We do not yet clearly understand the exact neural processes underlying the constructive aspect of memory. However, from the standpoint of the hippocampus, both memory and imagination may be manifestations of the same underlying neural processes. In the following chapters, we will explore the hippocampal neural processes underlying memory and imagination.

Chapter Three

PLACE CELLS AND HIPPOCAMPAL REPLAY

Human studies have shown that the same brain structure (the hippocampus) is involved in both memory and imagination, which may explain why we sometimes form memories of events we have never experienced (false memories). In this chapter, we will examine studies using animal subjects that provided important insights into hippocampal neural processes underlying memory and imagination.

HUMAN STUDIES VERSUS ANIMAL STUDIES

Two lines of research—one using human subjects and the other using animal subjects, rodents in particular—have driven the advancement of our knowledge of the hippocampus. Human studies yield data that is directly relevant to our ultimate interest: the functions of the human hippocampus and their underlying neural processes. However, it is difficult to interrogate hippocampal neural processes directly in humans because invasive approaches, such as recording neuronal activity with a microelectrode, are not allowed in humans. Such invasive studies are performed only rarely in human patients, and usually before brain surgery, to locate the source of epileptic seizures.

Animal studies are advantageous in this respect because we can perform a diverse array of invasive experiments targeting the hippocampus. For example, we can implant microelectrodes in the hippocampus and monitor in real time the spiking activities of individual neurons in a freely behaving animal.

We can also manipulate the activity of neurons in the hippocampus, such as by silencing a particular group of neurons and examining its effects on animal behavior. Such animal studies provide valuable information to our ultimate interest—understanding the functions and neural processes of the human hippocampus—because this evolutionarily old structure is remarkably similar in many respects between humans and other mammals. Therefore, human and animal studies on the hippocampus are complementary.

Animal studies have taught us a great deal about how the hippocampus stores memories. For example, the discovery of long-lasting changes in synaptic efficacy (i.e., changes in connection strengths between neurons; synapses are the connections between neurons) has greatly advanced our understanding of the neural processes underlying the hippocampal memory storage.[1] We can explain hippocampal memory formation mechanistically by activity-dependent changes in synaptic efficacy that lead to altered neural network dynamics (I will elaborate on this idea in chapter 4). We also have a good understanding of the molecular processes underlying synaptic efficacy change. It is now even possible to express synthetic proteins in specific hippocampal neurons and activate them to implant false memories into an animal.[2] As such, of all higher-order brain functions, neural mechanisms of memory are the best understood, largely due to findings from animal studies.

What about animal studies on the hippocampal role in imagination? Is it feasible to study this subject using animals? Definitely. Studies using rats during the last two decades yielded ample evidence that the rodent hippocampus is involved in remembering the past as well as imagining the future. Human and animal studies independently concluded that the hippocampus is involved in imagination as well as memory. More importantly, animal studies have provided significant insights into the neural circuit processes underlying the imagination function of the hippocampus that human studies cannot provide.

WHAT IS A REPLAY?

How did scientists discover the involvement of the rat hippocampus in imagination? By implanting microelectrodes and monitoring hippocampal neuronal activity in real time in freely behaving rats. Neurophysiologists implant microelectrodes in a certain region of the brain, monitor the activity of the neurons there, and then try to understand how that part of the brain processes information by analyzing neuronal activity in association with behavior. Using this approach, scientists found hippocampal "replays," which allowed them to understand the neural circuit processes underlying hippocampal imagination.

A hippocampal replay refers to the reactivation of a past experience-related neural activity sequence during an idle state. The sequence of neural activity observed during a rat's active navigation is repeated (or replayed) while the rat is sleeping or resting quietly. We will closely examine the hippocampal replay and how it might be related to the imagination in the rest of the chapter.

PLACE CELL

The most prominent characteristic of hippocampal neurons in freely moving rodents (e.g., rats and mice) is place-specific firing. Hippocampal neurons tend to be selectively active when an animal is within a restricted location of a recording arena. John O'Keefe, a neurophysiologist who reported this finding in rats with Jonathan Dostrovsky in 1971, referred to these neurons as "place cells."[3] With Lynn Nadel in 1978 he also published the influential *The Hippocampus as a Cognitive Map*, which considered the hippocampus the brain structure that represents the layout of the external space.[4]

Later studies have found place cells in the human hippocampus as well.[5] These place cells are thought to be the manifestation of the mental representation of the external space (or *cognitive map*). Some neurons in your hippocampus are probably active while you are sitting on a couch in your living room. As you stand up and walk toward the kitchen, different groups of place cells would then be activated in sequence at specific locations along your trajectory to the kitchen. O'Keefe was awarded a Nobel Prize in 2014 for the discovery of place cells that constitute "a positioning system, an 'inner GPS' in the brain that makes it possible to orient ourselves in space."[6]

PARALLEL TETRODE RECORDING

Early studies on place cells focused on how the external spatial layout is represented in the hippocampus. Even though an episodic memory involves a temporal element, only a few researchers attempted to understand how the hippocampus might store a sequence of events.[7] Thus, most early studies focused on hippocampal neural processes for representing static patterns rather than sequences of events. The discovery of hippocampal replays in the early 2000s changed this research trend dramatically. The team led by Matt Wilson at MIT discovered that sequential firing of hippocampal place cells during active navigation is repeated while rats are sleeping. This discovery set a new trend in research investigating the dynamics of place cells that represent spatial trajectories.

Matt and I worked together for four years in Bruce McNaughton's laboratory as postdocs at the University of Arizona. Matt developed a new recording technique that later enabled the discovery of hippocampal replays and I fortunately witnessed the entire development process as a colleague. Back then, neurophysiologists typically recorded one or at most a few neurons at a time. It was therefore difficult to monitor sequential activity patterns of multiple neurons, which is necessary to study the real-time dynamics of a neural network. Matt developed a new technique, a parallel tetrode recording system, and it allowed simultaneous recordings of up to a hundred hippocampal neurons in freely behaving rats. Rapid progress in research techniques now enables recordings from thousands of neurons simultaneously. In the 1990s, however, a simultaneous recording of up to a hundred neurons was a great leap forward that opened the door for investigating the real-time dynamics of a neural network. Matt continued this line of research and discovered hippocampal replays in the early 2000s.

THETA RHYTHMS, LARGE IRREGULAR ACTIVITY, AND SHARP-WAVE RIPPLES

To understand hippocampal replays better, let's first examine two different activity modes of the hippocampus. If you implant a microelectrode and monitor the electrical activity of a rat's hippocampus, you will easily identify two very different activity modes depending on behavioral states. The hippocampus shows strong rhythmic activity at around 6–10 Hz (theta rhythms) when the rat is active. In contrast, when the rat is resting, a slow and irregular rhythmic activity (large irregular activity) is observed along with occasional "sharp waves" (see fig. 3.1).[8]

Sharp waves represent massive synchronous discharges of many hippocampal neurons. Sharp waves occur along with fast oscillations known as "ripples" (140–200 Hz frequency); thus, they are commonly referred to as *sharp-wave ripples*. Notably, theta rhythms and sharp-wave ripples are also observed during sleep. Theta sleep in rats is thought to correspond to REM

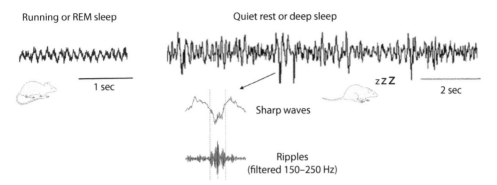

FIGURE 3.1. Two activity modes of the hippocampus. Theta rhythms (left) are observed during active movement and REM sleep, while large irregular activity (right) is observed during quiet rest and slow-wave (deep) sleep. Sharp-wave ripples are commonly observed during a period of large irregular activity. Sharp-wave ripples represent massive synchronous discharges of a large number of hippocampal neurons. Panels adapted from (left) Angel Nunez and Washington Buno, "The Theta Rhythm of the Hippocampus: From Neuronal and Circuit Mechanisms to Behavior," *Frontiers in Cellular Neuroscience* 15 (2021): 649262 (CC BY) and (right) Wiam Ramadan, Oxana Eschenko, and Susan J. Sara, "Hippocampal Sharp Wave/Ripples During Sleep for Consolidation of Associative Memory," *PLoS One* 4, no. 8 (August 2009): e6697 (CC BY).

sleep in humans, when most dreams occur, and brain activity is similar to waking levels. Sharp-wave ripples in large irregular activity are observed during slow-wave sleep in rats, which is thought to correspond to deep sleep in humans.

REPLAYS DURING REM SLEEP

Matt first investigated whether the sequential firing of place cells during active navigation is repeated during theta sleep in rats. This was based on the idea that dreams, which occur mostly during REM sleep in humans and, hence, probably during theta sleep in rats, may represent the process of transferring hippocampal memories elsewhere in the brain. As elaborated in chapter 1, the systems-level memory consolidation theory posits that new memory is rapidly stored in the hippocampus and then goes through a consolidation process so that it is eventually stored elsewhere in the brain.[9] Under this theory, it was natural to suspect dreaming as a process of reactivating memories stored in the hippocampus for systems-level consolidation. Indeed, consistent with this hypothesis, Matt and his student, Kenway Louie, found that sequential place-cell firing during active navigation is repeated during theta sleep. This landmark discovery was published in 2001.[10] The speed of sequential place-cell firing during theta sleep is similar to (in fact, slightly slower than) that during active navigation. This suggests that the progress of events in dreams may be similar to reality.

REPLAYS ASSOCIATED WITH SHARP-WAVE RIPPLES

REM sleep comprises about 25 percent of a night's sleep in humans. In rats, theta sleep represents less than 10 percent of the total sleep period, and the rest is mostly slow-wave sleep during which large irregular activity and sharp-wave ripples are observed. So what happens during slow-wave sleep? If the assumption is that memory consolidation takes place during theta sleep, then does no consolidation take place during slow-wave sleep? Does the brain simply take a rest without processing information? What happens during sharp-wave ripples when a large number of hippocampal neurons are synchronously active?

A follow-up study by another of Matt's students, Albert Lee, provided clear answers to these questions.[11] A close examination of hippocampal neural activity during sharp-wave ripples revealed that hippocampal neurons fire sequentially on a milliseconds time scale rather than in exact synchrony. More importantly, the sequence of hippocampal neuronal spikes during sharp-wave ripples is not random but is related to sequential firing during active navigation. Place cells tend to fire during sharp-wave ripples in the same sequence that they fired during active navigation before sleep (see fig. 3.2).[12] The speed of sequential firing is about fifty times faster during sharp-wave ripples than in actual navigation, suggesting that past navigation experiences are replayed in a time-compressed manner during sharp-wave ripples. This monumental discovery paved the way for subsequent research on the hippocampal neural circuit dynamics underlying memory, planning, and imagination. It was followed by a flourish of studies on sharp-wave ripple-associated replays in the rat hippocampus. Now, hippocampal replays generally indicate time-compressed reactivation of sequential hippocampal neuronal discharges that occur together with sharp-wave ripples.

I already mentioned that sharp waves occur during slow-wave sleep as well as during conscious quiet rest (see fig. 3.1). What happens then during sharp-wave ripples in the awake state? Do replays occur during awake sharp-wave ripples as well? Later studies demonstrated that hippocampal replays indeed occur in association with sharp-wave ripples during awake quiet rest as well.[13] Thus, regardless of state (sleep or awake), hippocampal replays appear to occur when sharp-wave ripples are observed.

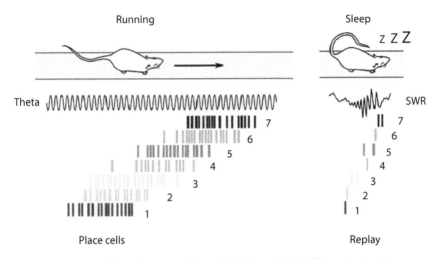

FIGURE 3.2. Sequential firing of hippocampal place cells during navigation (left) is replayed in a time-compressed manner in association with sharp-wave ripples (SWR) during slow-wave sleep (right). Each tick mark represents a spike. Figure adapted from Celine Drieu and Michael Zugaro, "Hippocampal Sequences During Exploration: Mechanisms and Functions," *Frontiers in Cellular Neuroscience* 13 (2019): 232 (CC BY).

Note that the hippocampal sharp-wave ripple is common and frequent. It occurs a few times every minute in humans and more frequently in rats during sleep and quiet rest.[14] In other words, replays of past experiences are likely to be ongoing while we rest or sleep. Considering that neuronal activity often leads to changes in neural network dynamics by inducing long-lasting changes in synaptic efficacy (referred to as activity-dependent synaptic plasticity), it is likely that hippocampal replays occurring during sharp-wave ripples have profound impacts on the representation of recently acquired memories. This suggests that memory consolidation might be ongoing not only during sleep but also during waking periods. That hippo-campal replays occur during the awake state also suggests that replays may serve functions other than memory consolidation, such as recalling past navigation trajectories (memory retrieval) and planning future navigation trajectories (planning the future).

REPLAY AND IMAGINATION

Early studies on hippocampal replays focused on the reactivation of previously experienced place-cell activity sequences. However, a study published in 2010 demonstrated that this is not the whole story.[15] David Redish and

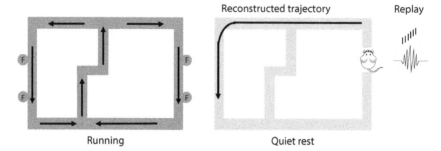

FIGURE 3.3. Replays for unexperienced trajectories. (Left) Arrows indicate the movement directions of a rat. Circles (F) represent food-delivering feeders. (Right) The rat's movement trajectory was reconstructed based on the sequential firing of place cells during a sharp-wave ripple while the rat was sitting quietly on the maze. Note that the rat never traveled along this sample reconstructed trajectory. Figure by author.

colleagues at Minnesota University trained hungry rats on a figure eight–shaped maze (see fig. 3.3) to alternatively visit the left and right pathways to obtain food pellets (see the arrows in the figure for the rat's spatial navigation trajectories). They then reconstructed the rat's spatial trajectories according to the sequential discharges of place cells (replays) during sharp-wave ripples. They found that some of the reconstructed trajectories matched those traveled by the rats. Importantly, they also found that some corresponded to trajectories the rats had never experienced before. In other words, hippocampal replays during sharp-wave ripples represent both experienced and unexperienced (but possible) spatial trajectories.

This study was a clear demonstration that hippocampal replays are more than mere reactivations of past experiences. Consider this finding together with what we examined in chapter 1 regarding the role of the hippocampus in imagination. This finding in rats fits well with the finding in humans that the hippocampus plays an important role in imagination. Two independent lines of research using humans and rats arrived at the same conclusion: the hippocampus is involved not only in memory but also in imagination.

REPLAYS IN HUMANS

So far, we have examined replays of sequential place-cell firing found in the rat hippocampus. But are these replays also found in humans? Does the hippocampus support replays of nonspatial sequences as well? Are replays coincident with the activation of the default mode network? Recent studies on humans provide positive answers to these questions.

It is difficult to measure sequential neural activity with the milliseconds time scale (replays) from deep brain regions (the hippocampus and related structures) using a noninvasive technique. Nevertheless, scientists were able to detect neural signatures of replay in the human brain by carefully analyzing brain signals measured by noninvasive techniques. Some commonly used techniques include electroencephalography (an EEG measures weak electrical signals), functional magnetic resonance imaging (an fMRI measures slow changes in blood flow; see chapter 1), and magnetic encephalography (an MEG measures weak magnetic signals from the brain). Each technique has its advantages and disadvantages. For example, MEG is advantageous over EEG for recording neural activity from deep brain structures (such as the medial temporal lobe) and over fMRI for temporal precision.

In recent studies using MEG, human subjects were presented with sequences of visual images. By carefully analyzing weak magnetic signals of the brain during a resting period, scientists were able to reconstruct correct stimulus sequences with a temporal resolution of 50 milliseconds. Notably, these replays were recorded together with the occurrence of sharp-wave ripples in the medial temporal lobe.[16] These human replays are cortical replays (recorded from the medial temporal cortex) rather than hippocampal replays. Nevertheless, considering that they were coincident with medial temporal lobe sharp-wave ripples and that cortical and hippocampal replays are found together in rats, these findings suggest common neural mechanisms for replay generation in rodents and humans.[17]

In another study using functional MRI, human subjects were presented with a sequence of images of houses and faces while performing a decision-making task. Scientists were able to reconstruct correct sequences on the order of 100 milliseconds based on the hemodynamic signals (changes in blood flow) measured from the hippocampus during conscious rest.[18] Together, these studies show that replays can be detected in the human brain using noninvasive approaches. Note that these human studies measured replays of nonspatial sequences. Thus, human hippocampal replays are not restricted to the spatial domain.

Notably, in the human subjects, the default mode network is active together with cortical replays and sharp-wave ripples in the medial temporal lobe.[19] This suggests that two independent findings from animal and human studies, namely hippocampal replay and default mode network activation, may be two sides of a coin. As such, findings from animal and human studies so far fit together nicely, indicating that hippocampal replays are the manifestation of neural processes underlying internal mentation, such as recalling the past and imagining the future, and occur during the activation of the default mode network.

SHARP-WAVE RIPPLE AS A GLOBAL PROCESS

Before wrapping up this chapter, let's examine what happens in the rest of the brain when a hippocampal sharp-wave ripple occurs. To investigate this matter, Nikos Logothetis and colleagues at the Marx-Plank Institute recorded activity of the monkey hippocampus with a microelectrode and monitored the global activity of the brain (regional blood flows) using a noninvasive technique (functional MRI) at the same time.[20] Surprisingly, almost all brain areas showed activity changes synchronized to the occurrence of a hippocampal sharp-wave ripple. Most cerebral cortical areas increased activity, while most subcortical areas decreased activity in association with sharp-wave ripples. For example, the thalamus, a subcortical structure that relays external sensory information to the cerebral cortex, decreases activity with the occurrence of a sharp-wave ripple. Moreover, even though most areas of the cerebral cortex increase activity, the primary visual cortex decreases its activity with a sharp-wave ripple, indicating that the brain disengages from external sensory signal processing at the time of a sharp-wave ripple.

Let's put this together with findings on the default mode network. Sharp-wave ripples tend to occur when our default mode network is active, i.e., when we are not paying attention to the external world. At this time, brain areas related to sensory processing (such as the primary visual cortex and thalamus) are silenced, and global information exchanges take place between the hippocampus and the cerebral cortex. Different brain areas appear to do their own jobs when we need to react to the external world, such as when chasing prey. However, when we have the opportunity to relax and can afford not to pay close attention to the outside world, our brains appear to switch off external sensory signal processing and instead allow global information exchange between different brain areas. This occurs when we are engaged in internal mentation, such as remembering the past, imagining the future, making moral decisions, thinking about the thoughts of others, and divergent thinking.

Neuroscience traditionally focused on how the brain processes information when a subject is actively responding to the external world. However, studies on the default mode network and replays suggest that these traditional approaches can tell us only part of the story about the way the brain operates. Recent findings indicate that to understand the principles of brain operation, we need to understand how different parts of the brain synchronize and exchange information during offline states and how these events affect signal processing during subsequent online states. The good news is that studies along this line are actively underway. New findings from these studies may change our traditional conceptual framework for studying the brain.

PART II

The Neural Symphony of Imagination

Chapter Four

NEURAL CIRCUITS OF THE HIPPOCAMPUS

In previous chapters, we examined the functional roles of the hippocampus, i.e., *what* the hippocampus does. In this chapter, we will explore *how* the hippocampus does what it does. Many neuroscientists are interested in a mechanistic understanding of brain functions in terms of neural circuit processes. Suppose you want to understand the mechanics of how a car uses gasoline when driving. A simple way to explain this would be that gasoline's combustive force allows the car to be driven. Would you be satisfied with this answer? Probably not. A more satisfactory answer would be that gasoline explosions inside a combustion engine force a piston to move up and down, a crankshaft translates the piston's linear motion into rotational motion, and the rotational power of the crankshaft turns the wheels of the vehicle. Similarly, neuroscientists are not satisfied with just knowing that the hippocampus serves memory and imagination functions. They want to go further to understand how the hippocampus serves these functions in terms of underlying neural circuit processes.

Thanks to the rapid technological advances of the past few decades, neuroscientists are now equipped with powerful tools to directly study the neural circuit processes underlying higher-order cognitive functions, such as decision-making and imagination. And considering the accelerating progress of technology, scientists are cautiously optimistic about the future of this endeavor—explaining higher-order cognitive functions in terms of underlying neural circuit processes (known as mechanistic cognitive neuroscience).

In this chapter, to build a foundation for exploring the neural circuit processes underlying memory and imagination functions of the hippocampus,
we will examine the anatomy of the hippocampus along with a classic theory
for the functioning of the hippocampal neural network.

TRISYNAPTIC CIRCUIT

The hippocampus is a sausage-shaped structure located deep in the medial
temporal lobe. Figure 4.1 shows a hand drawing of the hippocampal cross-
section by Santiago Ramón Cajal, a Spanish neuroanatomist and a Nobel
laureate (1909). The dentate gyrus, CA3, and CA1 are three major subregions of the hippocampus. The circuit that contains these structures (dentate
gyrus, CA3, and CA1) is known as the *hippocampal trisynaptic circuit*. CA
stands for cornu Ammonis, named after the Egyptian deity Amun, who has
the head of a ram. Neurons in the dentate gyrus project to CA3 and neurons
in CA3 in turn project to CA1.

This circuit (dentate gyrus → CA3 → CA1) has long been considered the
most prominent neural pathway of the hippocampus and, therefore, has
been the main target of research on hippocampal memory encoding and

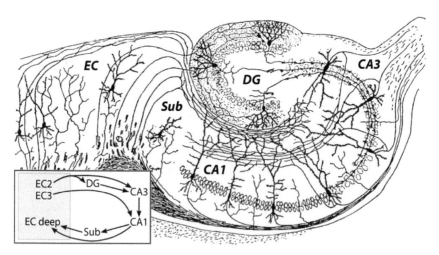

FIGURE 4.1. Santiago Ramon Cajal's hand drawing of the hippocampal cross-section. DG stands for dentate
gyrus, Sub for subiculum, and EC for entorhinal cortex, which is the main gateway of the hippocampus.
The inset shows a schematic of neural connections. EC2, EC3, and EC deep denote the entorhinal cortex's
layer 2, layer 3, and deep layers, respectively. Panel from "File:CajalHippocampus (modified).png," Wikimedia
Commons, updated April 19, 2008, https://commons.wikimedia.org/wiki/File:CajalHippocampus_(modified)
.png (CC BY).

retrieval. Note, however, that there are alternative connection pathways in the hippocampus (see inset of fig. 4.1). Moreover, recent studies indicate that a relatively small structure, CA2, has its own functions, such as social recognition memory (remembering events related to conspecifics).[1] However, because there are relatively few studies on CA2 and CA2's contribution to overall hippocampal functions is unclear, we will not examine CA2 here. Also, for the sake of simplicity, we will focus only on CA3 and CA1, leaving out the dentate gyrus, because they appear to be the core structures directly related to imagination, the main topic of this book. Brief discussions on the role of the dentate gyrus can be found in chapter 6 and appendix 1.

CA3 NETWORK

CA3 has long been at the center of the theoretical debate on the neural circuit processes that underpin hippocampal memory. It is also thought to be the central structure for the imagination function of the hippocampus. Why? Because CA3 neurons can excite each other. How? CA3 neurons not only project to other areas, such as CA1, but also back to themselves via massive *recurrent collaterals*, which are axon branches that ramify near the cell body (see fig. 4.2). Recurrent projections, which are commonly found

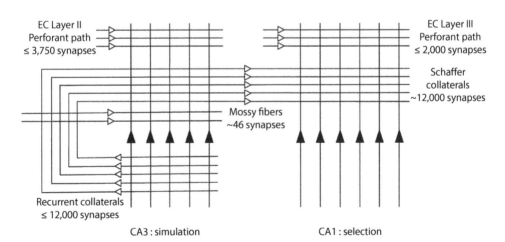

FIGURE 4.2. A schematic for CA3–CA1 neural circuits. Black triangles represent excitatory neurons and white triangles represent the projection directions of axons. EC stands for entorhinal cortex. Figure reproduced from Min W. Jung et al., "Remembering Rewarding Futures: A Simulation-Selection Model of the Hippocampus," *Hippocampus* 28, no. 12 (December 2018): 915 (CC BY).

in the cerebral cortex, allow internal interactions within a group of neurons. Such interactions enable a neural network to maintain and process information by internal dynamics in the absence of external input, a process thought to be essential for supporting higher-order cognitive functions of the cerebral cortex. Otherwise, a neural network will behave like a reflex machine responding simply to the presence of external input and be idle otherwise.

One defining characteristic of CA3 is an unusually large number of recurrent projections that connect its neurons. On average, one CA3 neuron receives about twelve thousand inputs from other CA3 neurons in the rat hippocampus. This amounts to three-quarters of all excitatory inputs a CA3 neuron receives (see fig. 4.2)![2] The sheer number of CA3 recurrent projections is indeed remarkable. What comes to your mind when you see such massive mutual connections? As you may have guessed, the CA3 neural network has a strong tendency for self-excitation. A side effect of such a neural network structure would be a tendency for a runaway excitation, i.e., seizures when network activity is not properly controlled. Epilepsy refers to a neurological disorder with repeated seizures. Considering the massive mutual connections among CA3 neurons, it is not surprising that temporal lobe epilepsy is the most common form of focal epilepsy. Henry Molaison (see chapter 1) suffered from such severe temporal lobe epilepsy that he underwent medial temporal lobe bisection.

CONTENT ADDRESSABLE MEMORY

Seizures are outcomes of abnormal circuit operations. What then is the physiological function of the CA3 neural network? Currently, the most influential theory on this issue is the one proposed by David Marr in 1971.[3] According to this theory, the CA3 neural network stores memories by changing the strengths of synaptic connections among CA3 neurons (i.e., by altering the synaptic efficacy of CA3 recurrent projections). More specifically, synaptic connections among CA3 neurons representing a certain experience are strengthened, forming a functional "cell assembly" of CA3 neurons representing that experience.[4] Later, if part of the functional assembly is activated, the rest of the CA3 neurons in the functional assembly are activated as well (because their connections are strengthened), therefore recovering the original activity pattern for the experience (memory retrieval). This process is explained schematically in figure 4.3.

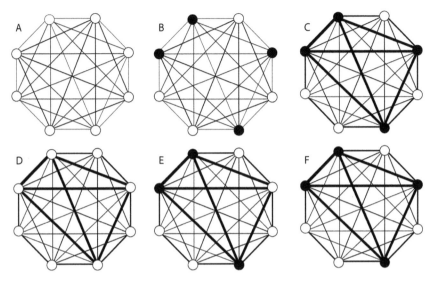

FIGURE 4.3. Storage and retrieval of memory in a neural network. Schematics illustrate how memory is stored in and retrieved from an interconnected neural network. Open and closed circles represent inactive and active neurons, respectively. Lines represent bidirectional connections between two neurons with the thickness indicating connection strength. The eight neurons in each neural network are all connected by recurrent projections. Let's assume that the network experiences an event, X. (A) shows the state of the network before experiencing X, and (B) occurs while experiencing X. We assume here that simultaneous activation of the four neurons (filled circles) corresponds to the perception of X. (C) shows that connections among simultaneously active neurons are strengthened to form a functional assembly. (D) occurs sometime after X. Although all neurons are inactive, enhanced connection strengths among the four neurons in the functional assembly persist, which is a long-lasting trace of memory. (E) shows a state of the network in response to a stimulus that is relevant to X. If the network did not experience X and no connection strength changes occurred, then the stimulus would be perceived as a new experience. (F) However, because synaptic connections among the four neurons of the functional assembly have been enhanced, the activity of three neurons will drive the activity of the one remaining. This way, the original pattern is completed, and the original memory is retrieved.

Scientists generally believe that a neural network stores and retrieves information in this manner, which differs from the way digital computers store and retrieve information. A digital computer stores information (memory) at a given address (e.g., a certain physical location on a hard disk) and retrieves the stored information by reading out information from that address. In contrast, a neural network stores information in a distributed manner by changing connection strengths among active neurons, as shown in figure 4.3. The stored information is retrieved by recovering the original activity pattern from a partial activation, a process called *pattern completion*. In other words, partial activation of the stored memory is necessary to retrieve the original activity pattern. This type of memory is sometimes

FIGURE 4.4. Memory retrieval by pattern completion. We can recall a memory by completing an original pattern from an incomplete or degraded input.

called *content-addressable memory* in the sense that the content of a particular memory is used to retrieve the memory, such as the memory address of a digital computer.

A memory comes to our mind when we encounter a related sensory input or while thinking about things related to that memory. We recall memory based on neural activity related to the stored memory rather than by exploring a certain location of the brain (address) where the memory is stored. Figure 4.4 illustrates this point. Suppose you encoded visual information on someone's face as a memory. Then you can recall the face easily when given partial or degraded visual information about the face.

SEQUENCE MEMORY

Marr's theory was adopted and further developed by later scientists. The core idea, that the CA3 neural network stores memory by enhancing connection strengths among coactive CA3 neurons, has been highly influential for the last fifty years. However, Marr's theory deals with the storage of static patterns, not sequences. When we examined hippocampal replays in chapter 3, we saw that the hippocampus does not merely store snapshots of our experiences but also stores event sequences that unfold over time. In addition, we now know that the hippocampus is involved not only with memory but also with imagination. Marr's theory is limited in explaining these aspects of hippocampal functions.

The discovery of hippocampal replays sparked interest in the neural processes underlying the storage and retrieval of experienced event sequences (episodic memory) and the generation of unexperienced event sequences (imagination). Again, CA3 is at center stage in this line of research. Because CA3 neurons are interconnected by massive recurrent projections, partial activation of CA3 neurons may sequentially activate other CA3 neurons. Physiological studies also showed that sharp-wave ripples, during which most hippocampal replays are detected, are initiated in CA3.[5] Therefore, there are good reasons to believe that CA3 plays a critical role in storing and retrieving experienced event sequences and in imagining unexperienced event sequences.

CA1 NETWORK

CA1, the final stage of the trisynaptic circuit, receives massive inputs from CA3 and sends outputs to surrounding areas of the hippocampus. In this sense, we may consider CA1 an output structure of the hippocampus. What then is the function of CA1? Even though numerous theories have been proposed regarding its role, empirical evidence is only tenuous, and nothing has been accepted as widely as Marr's theory on CA3. Anatomically, CA1 is different from CA3 in that it lacks strong recurrent projections;[6] the majority of CA1 neuronal axons are directed to the outside of the hippocampus. Thus, unlike CA3 neurons, CA1 neurons are likely to process information with little direct excitatory interaction among them. This suggests that neural network dynamics and functional roles of CA1 are likely to differ vastly from those of CA3.

You may then predict that neurophysiological characteristics, such as place-specific firing, would differ greatly between CA3 and CA1. Contrary to this prediction, scientists failed to find major neurophysiological differences between CA3 and CA1 neurons. Place cells are found in both CA3 and CA1 with largely similar characteristics. This suggests that the nature of processed information, particularly spatial information, is similar between CA3 and CA1. So why do two anatomically distinct structures process similar information? Why is CA1 even necessary if it processes similar information? We will examine this issue in detail in the next chapter. For now, my brief answer is that CA1 probably selects and reinforces high-value sequences generated by CA3.

I would like to show my respect to David Marr (1945–1980) before wrapping up this chapter. He made enormous contributions to the field of neuroscience. He proposed the first realistic neural network model for the hippocampus when our knowledge of the hippocampal neural networks was limited. More surprising is that his work on the hippocampus represents only a fraction of his contributions to neuroscience. He also proposed influential theories for the neocortex and cerebellum, and he is best known for his work on vision. Overall, Marr is considered a pioneer and founder of modern computational neuroscience. I think that hippocampal researchers can best show him respect by advancing his long-standing theory of the hippocampus. Perhaps all scientists, including him, would be happy to see their theories replaced by the next generation's work. This is how science advances.

Chapter Five

VALUE-BASED DECISION-MAKING

In this chapter, we will examine hippocampal value processing as the last piece of the puzzle before building a model of hippocampal neural circuit processes underlying memory and imagination. So how does value processing relate to hippocampal roles in memory and imagination? This will become clear in chapter 6 when we examine a simulation-selection model of the hippocampus. As a brief answer in advance, the model proposes that CA3 generates diverse activity sequences by self-excitation for both experienced and unexperienced events, and CA1 selects high-value activity sequences among them so that we can make better choices in the future. According to this model, the hippocampus is a device used to plan for future events rather than simply remembering what happened in the past. This model is grounded on the finding that the CA1 region of the hippocampus processes and represents value signals. We will examine this issue in detail in this chapter and, in doing so, briefly overview reinforcement learning and decision neuroscience.

VALUE, UTILITY, AND VALUE FUNCTION

Much progress has been made over the last twenty years in understanding the neural basis of value-based decision-making. A central assumption in studying the topic is that humans and animals make choices by representing values for potential choices. In the field of decision neuroscience, *value* refers to a long-term expected return. A closely related concept in economics is *utility*, which refers to the usefulness or enjoyment one can get from consuming a product or service. Suppose you need a smartphone, and you are considering two specific models with similar prices: an Apple iPhone and a Samsung Galaxy. The amount of subjective satisfaction one can get from each smartphone will vary from person to person. The iPhone may have a larger utility for you than the Galaxy phone—or perhaps the other way around. You will likely purchase the one with the largest utility. The current mainstream economics, neoclassical economics, is based on the assumption that people (*Homo economicus*) make choices to maximize their utility.

The idea of value in decision neuroscience is referred to as *value function* in reinforcement learning, which is a branch of artificial intelligence. Why is this branch of artificial intelligence known as *reinforcement learning*? Economics tries to explain and predict economic issues and problems by assuming that consumers and firms try to maximize their utility based on full and relevant information about their decisions. In contrast, reinforcement learning focuses on how a decision-maker finds precise value functions and an optimal choice strategy while interacting with an uncertain and dynamic environment. In other words, reinforcement learning is interested in the process of *learning* to approximate true value functions and finding an optimal choice strategy for maximizing long-term returns. Reinforcement learning laid the foundation for modern decision neuroscience by providing the major theoretical framework for neuroscientific investigation of decision-making.

REWARD PREDICTION ERROR

How does the brain process and represent value? It was not until the 2000s that this topic became popular among neuroscientists. Before this, neuroscientists tended to treat the brain like a computational device, focusing on the computational processes behind sensory processing and behavioral control. Few neuroscientists were interested in the neural process behind decision-making based on subjective values, such as choosing between a cappuccino and a latte, which would vary according to personal taste. In other words, neuroscientists used to be interested in *how* the brain processes sensory information and controls behavior rather than *why* the brain chooses one behavior over others. A dramatic change in this trend was triggered by a theoretical paper on dopamine neurons that was published in 1997.[1]

To get a clearer sense of the impact of this paper, we have to understand what *reward prediction error* is. As mentioned earlier, reinforcement learning deals with the process of finding precise value functions while interacting with an uncertain and dynamic environment. How? The simplest means would be through trial and error; you update your current value functions (expected rewards) according to the outcomes of your choices (actual rewards). As you iterate this process many times, your subjective value functions will converge to true value functions.

Suppose there are five restaurants to choose from for your lunch. Your value functions for them (the subjective satisfaction you can get from each restaurant) would be similar at the outset, but they will soon diverge as you visit them multiple times over a period. You increase your value function for a particular restaurant if a meal there was better than you expected and, conversely, decrease your value function if a meal was worse. This way, as you repeat your visits to these restaurants, your value functions for them will become closer to the true value functions. The world is not static, however. In our example, a particular restaurant may come up with a new menu or hire a new chef, which will make your current value function deviate from the true value function for the restaurant. Even so, if you update your value function every time you visit the restaurant, your value function will eventually catch up with the changes and approximate the new, true value function.

As you can see from this example, a key component for reinforcement learning is the difference between the actual and expected rewards (the reward prediction error). You increase your value function for a particular choice if the outcome is better than your current value function (when your actual reward is better than your expected reward, it's a positive reward prediction error). Conversely, you decrease your value function for that choice if the outcome is worse than your current value function (when your actual reward is worse than your expected reward, it's a negative reward prediction error). Of course, the value function is not modified if the actual and expected rewards are equal. This way, reinforcement learning can approximate true value functions in an uncertain and dynamic environment.

In their 1997 paper, Wolfram Schultz, Peter Dayan, and Reid Montague showed that midbrain dopamine neurons signal the reward prediction error.[2] These neurons, which project to widespread regions of the brain, play important roles in voluntary movements and reward processing. Abnormality in the dopaminergic neural system is associated with diverse neurological and mental disorders such as schizophrenia, Parkinson's disease, and addiction. As can be predicted from their roles in reward processing and addiction, some midbrain dopamine neurons elevate their activity in response to reward delivery and reward-predicting stimuli.[3] The researchers extended this finding by showing that some midbrain dopamine neurons are only responsive to an *unexpected* reward. Their activity is correlated with the difference between the actual and expected rewards rather than a reward per se.

Figure 5.1 shows the spiking activity of a sample dopamine neuron recorded from a monkey. In the initial phase, a reward (juice) was given

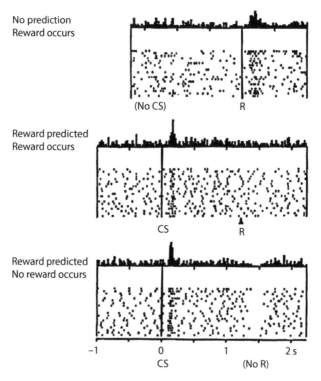

No prediction
Reward occurs

(No CS)   R

Reward predicted
Reward occurs

CS   R

Reward predicted
No reward occurs

-1   0   1   2 s
CS   (No R)

FIGURE 5.1. The activity of a sample dopamine neuron recorded from a monkey. This neuron increased activity when juice was given in the absence of a predicting stimulus (top row; R represents reward delivery). After pairing a sound cue (conditioned stimulus or CS) and juice delivery multiple times (the sound cue always preceded juice delivery), the dopamine neuron was not responsive to juice delivery anymore. The neuron was responsive to the sound cue (reward-predicting stimulus) instead (middle row). If juice delivery is omitted after the sound cue, dopamine neuronal activity is decreased at the time of expected juice delivery (bottom row). Thus, the activity of this neuron signals the difference between the actual and expected rewards (i.e., reward prediction error). Figure reproduced with permission from Wolfram Schultz, Peter Dayan, and P. Read Montague, "A Neural Substrate of Prediction and Reward," *Science* 275, no. 5306 (March 1997): 1594.

to the monkey sporadically so that the time of reward delivery could not be predicted. Per previous reports, this neuron elevated spiking activity in response to reward delivery (top row). In the next phase, reward delivery was always preceded by a sound cue one second before so that the time of reward delivery could be well predicted. At this phase, surprisingly, the dopamine neuron stopped elevating its activity in response to the juice delivery (middle row). In other words, this dopamine neuron did not respond to an expected reward. Furthermore, when juice delivery was omitted after the sound cue (i.e., expected reward is omitted), the dopamine neuron decreased its firing rate (bottom row). You can easily see that the magnitude of reward prediction error differs for the three conditions. Reward prediction error was positive when juice was delivered unexpectedly (sporadic reward delivery with no predicting stimulus; top row), zero when the same amount of juice was delivered as expected (reward delivery following the sound cue; middle row), and negative when juice was unexpectedly omitted (reward omission following the sound cue; bottom row). This and subsequent studies showed that dopamine neurons increase their activity if the actual reward is larger than expected and, on the contrary, decrease their activity if the actual reward is smaller than expected. This finding shows clearly that some midbrain dopamine neurons are responsive to reward prediction error—a key variable in reinforcement learning.

NEURAL REPRESENTATION OF VALUE

The finding that midbrain dopamine neurons convey reward prediction error signals suggests that the brain might represent values and update them according to actual choice outcomes. The next critical issue would be whether the brain represents value, another key component of reinforcement learning. The answer to this question is clear. Subsequent studies found value-related signals in many different parts of the brain in rats, monkeys, and humans. Neurons in these areas of the brain change their activity according to the type, amount, and probability of a reward.[4] Appendix 2 explains how scientists identify value-coding neurons with a specific example. These studies revealed that widespread regions of the brain represent reward prediction error and value, two critical elements of reinforcement learning. They also showed that choice behaviors of animals and humans are well accounted for by reinforcement learning algorithms in many behavioral settings.

The core idea of reinforcement learning is intuitive. You make choices according to your expectations and update your expectations according to actual outcomes. Your expectations will approach actual outcomes as you repeat this process, and your performance will improve accordingly. The findings of reward prediction error and value signals in the brain propelled further research on the neural basis of value-based decision-making. This line of research, decision neuroscience, is closely related to psychology, economics, and artificial intelligence. It is also called neuroeconomics since it studies the neural processes underlying economic decisions.

HIPPOCAMPUS AND VALUE

The hippocampus has long been considered a structure that represents cognitive signals, particularly spatial ones, rather than value signals; value-related information has been thought to be represented elsewhere in the brain. Recall the role of the hippocampus in declarative, but not procedural, memory (see chapter 1). Also, recall place cells and the role of the hippocampus in representing the spatial layout of the external environment (cognitive map; see chapter 3). Of course, it is well known that hippocampal neurons are responsive to reward (e.g., food) and punishment (e.g., an electric shock). However, the concepts of *reward* and *value* are not identical, even though they are related. Value is the final product of a cost-benefit analysis considering the type, amount, and probability of a reward along with the cost to obtain it. Value is the same concept as *expected return* in economics and finance, which is the profit an investor anticipates on a financial investment. Hippocampal responses to a reward are consistent with hippocampal value representation, but they may merely indicate the role of the hippocampus in representing experienced events (i.e., remembering the event of receiving a particular reward in a particular environment).

For these reasons, few scientists were interested in investigating value representation in the hippocampus. A growing body of findings indicates, however, that the hippocampus is among many brain areas involved in value representation. So far, value signals have been identified in the hippocampus of rats, monkeys, and humans.[5] Below, I will summarize a series of findings using rats in my laboratory that indicated hippocampal representation of value and hippocampal involvement in value-based decision-making. My laboratory, like many others, focused on the frontal cortex and basal ganglia in the initial exploration of the brain's value processing because they are targets of dopamine neuronal projections and are strongly implicated in reward-oriented behavior. One clear conclusion of this endeavor was that value processing is a ubiquitous function of the brain; we found value-responsive neurons in widespread areas of the rat brain.[6]

Studies in monkeys and humans also indicated that widespread areas of the brain are involved in value processing.[7] This suggests that value representation is an evolutionarily conserved function of the brain. It is often the case that the same function is found redundantly across many brain areas and sometimes across the entire body. For example, a biological clock is found in virtually all cells in our body, even though one master clock would be sufficient to control our circadian rhythms.[8] Evolution is a process of tinkering (modifying and improving an already existing system) rather than inventing an entirely new system from scratch. Hence, redundant functions are likely to be rooted deeply in evolution. As a rule, the more redundant a given function is, the more likely it is essential for survival and reproduction. Representing values of potential choices and making optimal choices based on them would be critical for survival and reproduction. It may not then be surprising to find value signals all over the brain.

As we found value-responsive neurons in various regions of the brain, we got curious about the possibility of value representation in the hippocampus. Because the hippocampus is an evolutionarily old structure, and because value-based decision-making would be critical for survival, we thought that the hippocampus might as well be involved in value processing. Hyunjung Lee, then a graduate student, investigated this matter by implanting micro-electrodes in the rat hippocampus.

VALUE REPRESENTATION IN CA1

Considering the traditional view that the hippocampus is specialized in representing spatial and cognitive signals, we initially thought that hippocampal value signals, if present, would be stronger in the output structures of the hippocampus rather than in the hippocampus itself. We therefore decided to examine neural activity in CA1, which is the final stage of the trisynaptic circuit, and the subiculum, which relays CA1 outputs to other cortical areas (see fig. 4.1). Our cautious prediction was that value signals would be found in the subiculum but not in CA1; or, if both areas convey value signals, the signals would be stronger in the subiculum than in CA1. This prediction turned out to be wrong.

Hyunjung trained thirsty rats to choose between two targets that delivered a water reward with different probabilities. The reward probabilities were not static but unpredictably changed over time. Thus, the rats had to decide which target to choose in a dynamic and uncertain environment. To maximize water intake in this circumstance, the rats had to figure out the reward probabilities of the two targets (values for the two target choices) based on the history of past choices and their outcomes and distribute their choices over two targets considering their relative values (i.e., reward probabilities). This process is well captured by reinforcement learning. Indeed, the rat's choice behavior in this task was well predicted by a simple reinforcement learning model, suggesting that the rats estimated and updated values for the two targets based on the history of past choices and their outcomes. Hyunjung estimated trial-by-trial values for the two targets using a reinforcement learning model and tested whether there are neurons whose trial-by-trial activity is correlated with value. See appendix 2 for a more detailed explanation of the procedure for finding value-coding neurons.

To our surprise, strong value signals were found in CA1 but not in the subiculum.[9] A large fraction of CA1 neurons, but only a small fraction of subiculum neurons, were responsive to value (a sample of value-coding CA1 neurons is shown in appendix 2). We were surprised but at the same time excited by this unexpected finding. A perplexing yet thrilling moment for a scientist occurs when facing an unexpected result. I began my neuroscience career in 1986 as a graduate student at the University of California, Irvine. I had thought about neural circuit processes underlying hippocampal functions for a long time but could not come up with a satisfactory answer, especially regarding the role of CA1 in hippocampal functioning. But this unexpected finding—that CA1 represents strong value signals—provided a breakthrough; we were able to devise a new model of the hippocampus, which I think captures the essence of hippocampal circuit operations, based on this finding. This model will be elaborated on in chapter 6.

VALUE REPRESENTATION IN CA3

We were shocked by the finding that CA1, rather than the subiculum, represents strong value signals. Why does the hippocampus, which is known to be primarily concerned with cognitive information, represent value signals? How is hippocampal cognitive signal processing affected by value signals?

What is the role of value signals in the functioning of the hippocampus? As a step to investigate these matters, we examined whether CA3, the major input structure of CA1, also represents value signals. Sung-Hyun Lee, then a graduate student, compared the value signals of CA3 and CA1.[10] The conclusion was clear. Value signals were much weaker in CA3 than in CA1. This result indicates that CA3 is not the source of CA1 value signals. Also, the fact that value signals are strong in CA1 but weak in its main input (CA3) and output (subiculum) structures suggests that value representation may be a special characteristic of CA1 among the hippocampal subregions.

WHAT IF CA1 IS INACTIVATED?

The results so far are outcomes of correlational studies that show spiking activities of many CA1 neurons are correlated with values of two targets. In general, correlational and interventional studies complement each other. We activate or suppress a certain part of the brain and examine its consequences on behavior in an interventional study. We can conclude with some confidence that a certain brain structure plays an essential role in a certain function when both correlational and interventional approaches yield converging results. In our case, that CA1 processes strong value signals (correlational approach) and CA1 inactivation impairs value-based decision-making (interventional approach) would make a strong case for the involvement of CA1 in value processing. Of course, correlational and interventional approaches do not always yield converging results. Even though strong neural signals related to a certain function were found in a brain structure, its inactivation may have no behavioral consequence. This is because a given function may be served redundantly by multiple brain structures.

With this background, we examined the behavioral consequences of inactivating the CA1 of mice. This work was done by Yeongseok Jeong, then a graduate student. In fact, I was somewhat reluctant to perform this study. As mentioned earlier, value signals are found in widespread areas of the brain. Moreover, numerous studies indicated the importance of the frontal cortex and basal ganglia in value-based decision-making. Thus, it seemed that hippocampus-inactivated animals may well rely on these neural systems for value-based decision-making. Moreover, many studies indicated the role of the hippocampus in the rapid encoding of declarative memory (see chapter 1) rather than incremental value learning, which was believed to be mediated by the basal ganglia.[11] For these reasons, I predicted that

CA1 inactivation would minimally affect a rat's choice behavior in tasks that require trial-by-trial adjustments of values according to trial outcomes. This experiment was nevertheless necessary to follow up on the finding that CA1 represents strong value signals. This is a type of study a graduate student may not be very enthusiastic about. It is generally difficult to publish a negative finding (i.e., CA1 inactivation does not affect behavior). In our case, if we found no effect of CA1 inactivation on the animal's choice behavior, it would be difficult to tell whether the absence of inactivation effect was because CA1 is dispensable for value-based decision-making or because CA1 was insufficiently inactivated. Thankfully, even with this caveat, Yeongseok agreed to perform this study with little hesitation.

My prediction turned out to be wrong one more time. Yeongseok found that CA1 inactivation alters the mice's choice behavior so that they were less successful in obtaining water.[12] He used a technique known as *chemogenetics* to selectively inactivate different subregions of the hippocampus (CA1, CA2, CA3, and dentate gyrus) and examined the mice's choice behavior.[13] He then used a reinforcement learning model to study which aspect of value-based decision-making was affected. He found that CA1 inactivation impairs value learning; the process of value updating based on choice outcomes (i.e., value updating based on reward prediction error) was compromised in CA1-inactivated mice. By contrast, the inactivation of CA2, CA3, or dentate gyrus had no significant effect on the mice's choice behavior whatsoever. I was once again surprised by this finding. In my estimate, the chance of getting a positive finding was well below 50 percent. We had some trouble publishing this finding in a peer-reviewed journal. This was probably because of the long-standing view in the field that the hippocampus is important for rapid encoding of facts and events but not for gradual value learning.

CA1: A VALUE SPECIALIST

Let's summarize the findings in rats and mice. First, CA1 represents strong value signals, but its main input and output structures, CA3 and subiculum, respectively, represent values only weakly. Second, CA1 inactivation impairs value-based decision-making, but dentate gyrus, CA2, or CA3 inactivation had no significant effect on the animal's choice behavior. The message of these findings is clear. Value representation, as a unique characteristic of CA1 among hippocampal subregions, is likely to play a crucial role in CA1 functioning.

The hippocampus has traditionally been thought to process cognitive information, especially spatial information. From this standpoint, the function of CA1 has been puzzling because spatial firing patterns are similar between CA3 and CA1 neurons. Of course, subtle differences are found between CA3 and CA1 place-cell characteristics, but overall, they are quite similar. As to the functional differentiation between CA3 and CA1, do both CA3 and CA1 represent spatial information (cognitive maps)? If so, why do we need both CA3 and CA1? Isn't CA3 alone sufficient to represent external spatial information? There exist numerous theories on this issue, but, to me, none is persuasive. Our findings provide a new perspective on this long-standing issue. That CA1 is a value specialist among all hippocampal subregions provides a clue to understanding the functional role of CA1.

Up to now, we have examined pieces of the puzzle to understand how the hippocampal neural network supports memory and imagination. In chapter 6, we will try to put them together to see the overall puzzle.

Chapter Six

REMEMBERING REWARDING FUTURES

Many studies targeting the hippocampus have been carried out. As of April 2022, typing in the word "hippocampus" in PubMed (a search engine for biological and medical publications) generates over 170,000 publications. These studies have contributed greatly to our understanding of the hippocampus. We now know a great deal about its anatomy, biochemistry, physiology, development, function, and relevance to neurological and mental diseases. I selectively introduced a few of these studies in the previous chapters that I think are critical for understanding the neural circuit operations underlying the memory and imagination functions of the hippocampus. To summarize, the hippocampus is involved not only in remembering past experiences but also in imagining future events; sharp-wave ripples and hippocampal replays take place in the hippocampus during sleep and resting states; CA3, but not CA1, has massive recurrent projections that enable self-excitation and sequential firing; CA3 generates sharp-wave ripples and CA1 represents value signals. We will examine a synthesis of these findings, a simulation-selection model, in this chapter.

SIMULATION-SELECTION MODEL

The main idea of the model is simple: CA3 generates diverse event sequences based on massive recurrent projections during rest and sleep (simulation) and CA1 preferentially reinforces high-value sequences based on value-dependent neural activity (selection). This way, neural activity sequences representing high-value events and actions will be preferentially reinforced so that they are likely to be chosen in the future under similar circumstances. This will allow us to make better choices in the future.

Together with my colleagues, I proposed this model in 2018.[1] We proposed that the hippocampus simulates and reinforces high-value events and actions in preparation for the future rather than merely remembering what happened in the past (the idea for a hippocampal role in future planning has been proposed by numerous scientists; see our paper and references therein). We also put forward that this function of the hippocampus is implemented in the CA3-CA1 network. This may look inefficient at first glance because not just one but two neural networks are needed to prepare for optimal choices. However, two networks separately performing simulation (CA3) and selection (CA1) enable the generation and evaluation of a wide variety of events and actions, which would be useful to prepare in advance for diverse future circumstances. If we rely on only one network for both simulation and selection, the diversity of simulated events or behaviors will be markedly reduced. We elaborated in our paper why we consider the hippocampus as a simulation-selection device and how the process of simulation-selection might be implemented in the CA3-CA1 neural circuits. Below, I will summarize the key arguments for the model.

CA3 AS A SIMULATOR

Why do we consider CA3 a simulator? The hippocampus plays a role in imagination (see chapter 1). It also generates place-cell firing sequences that correspond to unexperienced spatial trajectories during rest and sleep (see chapter 3). These findings indicate that the hippocampus generates novel activity sequences. Put differently, it simulates unexperienced event sequences.

Where then in the hippocampus are simulated sequences generated? Most scientists would point at the CA3 as the source because CA3 neurons are connected by massive recurrent projections (see fig. 4.2). CA1, in contrast, has only weak, longitudinally-directed recurrent projections.[2] Because CA3 neurons are heavily interconnected, activation of some CA3 neurons will likely activate others (self-excitation). Propagation of such sequential activation fits well with the sequential place-cell firing that occurs with a sharp-wave ripple during sleep and resting states (i.e., replays). As mentioned in chapter 4, sharp-wave ripples are initiated in CA3 and propagate to CA1.[3] Together, these findings consistently indicate CA3 is the source of simulated sequences.

How then does CA3 generate novel activity sequences during sharp-wave ripples? Why don't they simply repeat firing sequences that happened during past active states? Scientists believe that CA3 stores memories of experienced events, such as navigation trajectories, by changing connection strengths among CA3 neurons (see fig. 4.3). However, the following factors would act against repeating the same activity sequences exactly as during active navigation under resting or sleep states. First, individual synaptic communication is unreliable among brain cells because of the probabilistic release of neurotransmitters. A message from one neuron is transmitted to another neuron only probabilistically. Second, the brain state is likely to differ drastically between active navigation and passive resting states. In rats, theta-frequency rhythmic oscillations are dominant during active navigation, but slow oscillations and sharp waves are dominant during passive states (see fig. 3.1). Third, inhibitory neuronal activity is lower during passive compared to active states. Thus, it appears that the CA3 neural network is more loosely controlled under passive states. Fourth, incoming sensory inputs may differ drastically between active navigation and resting or sleep states. Finally, CA3 is a network interconnected with many individually weak synapses rather than a few strong ones. An unusual feature of the CA3 network is the sheer number of recurrent projections. As we examined in chapter 4, each excitatory CA3 neuron receives synaptic inputs from about twelve thousand other excitatory CA3 neurons, which comprises about 75 percent of all synaptic inputs it receives (see fig. 4.2). However, physiological studies have shown that recurrent projection synapses are individually weak, which would be disadvantageous for the generation of high-fidelity activity sequences.

To summarize, because CA3 is a network interconnected with many weak synapses, the CA3 neural network state differs greatly between active and passive states, incoming sensory inputs differ drastically between active and sleep states, and inhibitory regulation is weak during passive states, it would be difficult to repeat the same firing sequences as during active navigation under inactive states. Consequently, CA3-generated replays will consist of not only experienced but also unexperienced sequences. In this respect, randomness may be a critical functional element of the CA3 network. It would allow the network to generate a wide variety of unexperienced (novel) sequences and function as a simulator rather than a high-fidelity memory device.[4]

CA1 AS A VALUE-DEPENDENT SELECTOR

What will happen to replays that are generated in the CA3? Some researchers have proposed that hippocampal replays will be evaluated in brain structures such as the ventral striatum and orbitofrontal cortex, which are well known to process value-related signals.[5] This proposal is in line with the long-standing view that the hippocampus mainly processes spatial and cognitive information rather than value-related information. However, our results indicate that CA1 is a value specialist. A corollary of strong value-related CA1 neural activity is that CA3-originated neural signals will be processed differently in CA1 according to their associated values. In other words, CA1 will filter CA3-generated replays according to their associated values. For example, CA1 may preferentially pass high-value replays, such as those corresponding to spatial trajectories leading to a rewarding location, while filtering out low-value replays, such as those corresponding to spatial trajectories leading to an unrewarding location. Of course, "selection" and "filtering out" here by no means indicate an all-or-none process. CA3 replays will be more likely to pass through CA1 as their associated values increase. CA3 presumably generates a huge number of replays during resting and sleep states. We propose that CA1 processes these CA3-generated replays in proportion to their associated values so that high-value replays are preferentially selected and reinforced.

To understand how exactly the process of simulation-selection operates in the CA3-CA1 neural circuit, we need to compare how CA3 and CA1 replays are affected by their associated values. Few studies have explored this issue, but findings so far are consistent with the simulation-selection model. For example, CA1 place cells with their firing fields near a rewarding location are preferentially reactivated during sharp-wave ripples compared to those with their firing fields far from a rewarding location.[6] In contrast, CA3 place cells do not show such reward-dependent activation during sharp-wave ripples.[7] No other studies have compared the reward or value dependence of CA3 and CA1 replays so far. Nevertheless, numerous studies have repeatedly shown that reward facilitates CA1 replays in rats.[8] In humans, the imagination of episodic future events is enhanced by reward, and hippocampal activity patterns for high-reward contexts are preferentially reactivated during post-learning rest.[9] These results are well in line with the proposal that CA3 generates replays independent of their values while CA1 preferentially processes high-value replays. The functional consequence of this operation is clear. The selection of high-value sequences will strengthen neural representations for those sequences, which can guide optimal choices in the future.

DENTATE GYRUS

We have examined key concepts of the simulation-selection model. More of its details, especially those related to the neurobiological implementation of the simulation-selection process, can be found in the paper I published with my colleagues in 2018.[10] We focused on CA3 and CA1, leaving out the dentate gyrus, which is another component of the hippocampal trisynaptic circuit. What does the dentate gyrus do in hippocampal functioning? And how is its function related to the proposed simulation-selection process of the CA3-CA1 network?

Currently, *pattern separation* is the most popular theory for the role of the dentate gyrus. This idea is related to David Marr's theory that CA3 stores associative memory (fig. 4.3). The main thrust is that the dentate gyrus separates similar input patterns into distinct patterns so that CA3 can store many patterns (memories) with minimal interference.[11] However, it is unclear whether this idea can be applied to memories for sequences rather than static patterns.

There are other reasons to doubt that pattern separation is its major function. I have proposed, together with my long-term colleague, Jong Won Lee, that the primary function of the dentate gyrus is to bind together diverse sensory signals and, by doing so, form "spatial context."[12] Appendix 1 also briefly discusses this matter. To put it simply, we think that the trisynaptic circuit of the hippocampus performs what we call *binding-simulation-selection*. The dentate gyrus allows us to recognize where we are (our spatial context) by binding together diverse sensory signals, and CA3 and CA1 together perform simulation-selection to reinforce high-value sequences in each spatial context.

IMPLICATIONS OF THE MODEL

The simulation-selection model is a theory that awaits empirical verification. Nevertheless, it coherently explains findings that cannot be readily accounted for by conventional theories. For example, the model explains why the hippocampus is involved not only in memory but also in imagination, why memory is prone to falsification, why the hippocampus represents value, why the hippocampus needs CA1 in addition to CA3, and why place cell characteristics are similar across CA3 and CA1, in terms of a simple scheme of simulation-selection. In addition, the model provides new perspectives on neural processes underlying goal-directed behavior and memory consolidation.

First, the model explains two core processes of goal-directed spatial navigation, namely spatial and value representations, with a single neural mechanism. It is generally assumed that goal-directed spatial navigation is supported by spatial information represented in the hippocampus and value information represented elsewhere in the brain. However, both spatial and value information are represented in the hippocampus in the simulation-selection model; therefore, goal-directed spatial navigation can be explained by a simple process of simulation-selection within the hippocampus. There is no need to assume two separate neural systems dedicated to navigation and value processing.

Second, the model provides a new perspective on memory consolidation. We examined issues and debates on memory consolidation in chapter 1. It is still unclear why and how initially formed memories are consolidated over time to become permanent memories. The simulation-selection model posits that memory consolidation is a process of finding optimal strategies based on past experiences rather than strengthening incidental memories. This view is radically different from conventional theories on memory consolidation.

DYNA

Memory consolidation as a process of actively selecting and reinforcing valuable options for the future is surprisingly akin to a well-known machine learning algorithm. As mentioned in chapter 5, reinforcement learning is a branch of artificial intelligence that aims to find optimal action plans in a dynamic and uncertain environment. An agent selects actions based on value functions and updates value functions based on the consequences of actions. This iterative process allows an agent to keep track of true value functions and make adaptive choices.

One drawback of such a trial-and-error approach, however, is inefficiency. It often requires an enormous number of trials to approximate true value functions. This is particularly problematic when a long sequence of actions is needed to reach the final goal. It would be difficult to know whether selecting a particular action (X) in a situation (Y) is of high value or low value if the consequence of choosing that action in that situation is revealed only after a long sequence of actions. This explains why reinforcement learning algorithms in general have trouble mastering the video game *Montezuma's Revenge*, in which the character Panama Joe must go through many steps before getting to the destination, the Treasure Chamber, in Montezuma's pyramid.

One proposed solution to overcome this difficulty is to perform simulations during the offline state to supplement trial-and-error value learning. Imagine a robot vacuum cleaner trying to master an effective way to clean a room filled with furniture. Finding the best strategy to clean the room may take a long time if the arrangement of the furniture is complex because there would be an immense number of possible trajectories for covering the entire floor. If the robot vacuum cleaner relies solely on its actual cleaning experiences, it may take months, or even years, to figure out the best strategy for the room.

One way to solve this problem is to learn value functions by simulating possible trajectories. This is the core idea of the Dyna algorithm David Sutton proposed in 1991.[13] The algorithm learns value functions in two steps: it first learns value functions while interacting with the environment (trial-and-error learning), and it then learns value functions by simulating actions and assessing their outcomes (offline learning). In our example, the robot vacuum cleaner learns value functions for various spatial trajectories by actual cleaning and then by simulation without actual movement. This can greatly facilitate the rate of learning because the robot can evaluate an enormous number of trials without actually performing the cleaning process.

The similarity between the simulation-selection model and the Dyna algorithm is remarkable. Both increase the rate of value learning by simulation based on limited experiences. The process of representing accurate value functions in an uncertain environment may take a long time if we rely only on trial-and-error learning. Of course, we can eventually learn accurate value functions if an environment is stable. However, your competitors, such as potential predators, are not nice enough to wait for you until you represent value functions accurately. It's a jungle out there. Moreover, environments often change dynamically. With slow learning, you may never make optimal choices in a dynamic environment because the environment (and hence true value functions) may change before you master them. Let's assume that it takes one full year for the robot to master the best cleaning strategy for a room by trial and error. Let's also assume that the room's residents and furniture arrangement change every three months. If so, the robot will never be able to clean the room most efficiently. This problem can be solved by using simulation-selection to accelerate learning.

Chapter Seven

THE EVOLUTION OF IMAGINATION

Biological phenomena can be explained by proximal as well as ultimate causes. Proximal explanations concern underlying biological mechanisms, whereas ultimate explanations concern evolutionary origins and functional utilities. For instance, why do we get thirsty after eating salty food? One possible answer could be that a high salt intake increases blood osmolality, which is detected by osmolality sensors in the blood vessels that then activate the neural circuit that triggers water-seeking behavior (proximal cause). Another possible response could be that the biological machinery that detects and lowers high salt concentration has evolved because it is necessary for survival (ultimate cause). Our explanations of the simulation-selection model so far have focused on the proximal cause—the neural mechanisms of simulation-selection. What then is the ultimate explanation for the simulation-selection function of the hippocampus? Why has it evolved to perform simulation-selection? In this chapter, we will try to find an answer to this question by comparing the anatomy and physiology of the hippocampus across different animal species.

MAMMALIAN VERSUS AVIAN HIPPOCAMPUS

To recapitulate, replays found in the rat hippocampus suggest that it simulates diverse spatial trajectories during rest and sleep. Additionally, value-dependent activity of CA1 neurons suggests that CA3-generated spatial trajectories are selected in CA1 according to their associated values. Together, these findings suggest that the rat's hippocampus performs simulation-selection of diverse spatial trajectories, both experienced and unexperienced, during rest and sleep so that the rat can choose the optimal spatial trajectory between two arbitrary locations in the future.

What then is the ultimate cause of simulation-selection? My first question regarding this issue was whether the avian hippocampus also performs simulation-selection. This is because the bird, which can fly, may not need to exert effort to prepare for potential spatial trajectories in advance. The bird may travel directly "as the crow flies" from its current position to its destination.

Unfortunately, there are few studies on the bird hippocampus compared to many on the rodent hippocampus. The most popular animal models for the hippocampus are rats and mice. Using the same animal model allows researchers to compare results from different studies, which facilitates a more complete understanding of a biological phenomenon or function of interest. Moreover, because evolution is generally a process of modifying existing biological machinery rather than inventing something entirely new from scratch, findings from one animal species are likely to be applicable to another. For example, biological functions of a specific gene in the fruit fly, a widely used animal model in biology, are often directly related to human genetic diseases. However, there is also a need for comparative studies, which allow us to understand how general a particular biological mechanism is across species and how it varies according to the species' ecological needs.

The structure of the hippocampus is surprisingly similar across different species of mammals, including humans.[1] As shown by the brain cross sections of several mammalian species (see fig. 7.1), all mammalian hippocampi have similar gross anatomical structures with clearly distinguishable dentate gyrus, CA3, and CA1. In contrast, as shown by brain cross sections of nonmammalian vertebrates (fish, reptile, and bird; see fig. 7.2), these animal species do not have clearly defined hippocampal subregions that correspond to the dentate gyrus, CA3, and CA1 of the mammalian hippocampus. Anatomical studies have identified additional outstanding differences between the mammalian and avian hippocampus.[2]

Mammals and birds, warm-blooded vertebrates, have far better spatial memory capacity compared to cold-blooded vertebrates (fish, amphibians, and reptiles). This may be quite ironic considering that calling someone a "birdbrain" is considered an insult. In fact, many food-caching bird species have surprisingly superb spatial memory. Of a total of about 170 bird families,

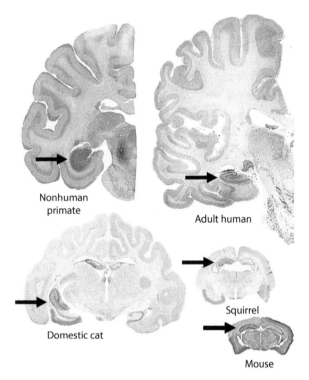

FIGURE 7.1. Brain cross sections of several mammalian species. Hippocampal cross sections (indicated by arrows) show clearly distinguishable dentate gyrus, CA3, and CA1 in all species. Figure adapted from Jaafar Basma et al., "The Evolutionary Development of the Brain as It Pertains to Neurosurgery," *Cureus* 12, no. 1 (January 2020): e6748 (CC BY).

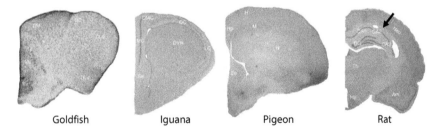

FIGURE 7.2. Brain cross sections of the goldfish, iguana, pigeon, and rat. The distinct cross-sectional structure of the mammalian hippocampus (arrow) is not shown in nonmammalian species (courtesy of Verner Bingman).

12 store food. For example, Clark's nutcrackers cache food, such as pine nuts, at thousands of locations and retrieve most of them in the winter.[3] Because these locations become snow-covered in the winter, they must remember them on a spatial map of the food-caching area rather than remembering local features unique to each spot. Although the hippocampus oversees spatial memory in both mammals and birds, the mammalian and avian hippocampi differ drastically in their anatomical structures. This tells us that the way the hippocampus handles spatial information may differ drastically between the two groups.

PLACE CELLS IN PIGEONS

A team led by Verner Bingman at Bowling Green University in Ohio implanted microelectrodes in the pigeon hippocampus to examine spatial firing of avian hippocampal neurons. Due to technical issues, they recorded neural activity while the pigeon was walking rather than flying. Nevertheless, their studies provided rare and valuable data on spatial firing of hippocampal neurons. Does the pigeon hippocampus have place cells like the mammalian hippocampus does? The short answer is no.[4] They trained pigeons to walk around in a plus maze to obtain food at the end of each arm. Interestingly, spatial activity patterns differed between the left and right hippocampi. Some neurons in the right hippocampus were active preferentially at many rewarding locations (known as *goal cells*) and some in the left hippocampus were active preferentially along the paths connecting two rewarding locations (*path cells*). This suggests that the pigeon hippocampus may only be concerned with the final target locations (i.e., rewarding locations) and the direct paths between them.

Another interesting finding from this research is that these spatial firing patterns diminish greatly if food rewards are scattered randomly in the maze. By contrast, rat hippocampal neurons show place-specific firing regardless of whether food rewards are scattered randomly or provided only at specific locations. These findings further indicate the difference between mammals and birds in the way the hippocampus processes and represents spatial information.

We can generate an infinite number of spatial trajectories by sequentially arranging place cells found in the rodent hippocampus. In contrast, it would be difficult to construct spatial trajectories based on the sequential firing of spatial neurons found in the pigeon's hippocampus. This casts a doubt on the implementation of simulation-selection in the pigeon's hippocampus.

EVOLUTION OF SIMULATION-SELECTION

Figure 7.3 illustrates the key point of our proposal that the simulation-selection function of the hippocampus has evolved in land-navigating mammals because they need to choose optimal trajectories between two arbitrary locations. This need, however, does not apply to birds; they can fly directly to a target location. Land-navigating animals often face obstacles, such as rivers, trees, and rocks, that block the straight path toward a target location. They, therefore, need to remember trajectories that get around these obstacles to get to a target location in an efficient manner. However, it would require too much time and effort to learn all the optimal trajectories

FIGURE 7.3. Goal-oriented navigation in mammals and birds. Land-navigating mammals, but not birds, need to remember spatial trajectories. Figure reproduced with permission from Min W. Jung et al., "Cover image," *Hippocampus* 28, no. 12 (December 2018).

between two arbitrary locations in each environment by experience. One way to resolve this dilemma would be to learn optimal trajectories without actual navigation, i.e., performing simulation-selection. That would explain why simulation-selection may have evolved in land-navigating mammals but not in birds.

One might argue that it would be more efficient to compute the optimal trajectory as necessary rather than spending time and energy in advance to figure out an enormous number of trajectories that may or may not be used in the future. Such a strategy, however, may be fatal in emergency situations, even though it would be efficient in terms of energy expenditure. For instance, let's assume that you are a rabbit grazing on a field away from your burrow. When you realize that a fox is charging toward you, you must run back home using the shortest available route. It may take too long to compute the optimal route in such a circumstance. A delay of one or two seconds may cost you your life. Advanced simulation-selection would prepare you to identify optimal navigational routes between an arbitrary starting location and your burrow. Hence, even though it may be time- and energy-consuming, it may be advantageous for survival to represent optimal trajectories from arbitrary starting locations to a few target locations in each environment. Furthermore, the environment may change dynamically. Lush and dense grass that used to block your passage may disappear in the winter, or a deep-water pit may appear on your favorite route after a heavy rain. It would then be advantageous for survival to keep updating your optimal strategies. Otherwise, a land-navigating species may not be able to survive in the long run.

Assuming so, when would be the best time to perform simulation-selection? Obviously, not when you are being chased by a predator. It would be more reasonable to perform it during rest or sleep when you don't need to pay attention to the outside world. This may be why the hippocampus, as a part of the default mode network, is active during resting states. Considering that the hippocampus is an evolutionarily old structure, and that it is globally synchronized with the rest of the brain during sharp-wave ripples, the process of simulation-selection may have been a crucial factor in the evolution of the default mode network. As we examined in chapter 1, the default mode network is active in association with various types of internal mentation, such as autobiographical memory recollection, daydreaming, moral decision-making, theory of mind, and divergent thinking; simulation is thought to underlie all these internal mentation processes.[5]

WHALES AND BATS

Considering my argument that the simulation-selection function of the hippocampus may have evolved in mammals because of the unique navigational need of land mammals, it would be interesting to examine the hippocampus of the mammals that abandoned ground navigation. Cetaceans, such as whales, went back to the sea and bats acquired wings that enable them to fly. These animals may have lower demands to store a large number of potential navigation trajectories compared to land-navigating mammals. How developed is the spatial firing of hippocampal neurons in cetaceans? To my knowledge, no one has recorded any data on that. But there are studies examining the anatomy of the whale hippocampus. These studies found that its relative size is much smaller compared to other mammals, suggesting a degeneration of the whale hippocampus during evolution.[6] This finding supports the possibility that the hippocampus of land-navigating mammals has evolved to find and remember optimal spatial trajectories between two arbitrary locations that can be used in the future. In comparison, the whale would rarely face obstacles during travel.

On the other hand, bats have well-developed hippocampi. Furthermore, physiological studies have found three-dimensional place cells (or *space cells*) in their hippocampi.[7] This does not support our ultimate explanation for the simulation-selection function of the hippocampus. Why do bats have clear place cells but pigeons don't, even though they can fly? It may be that the bat's lifestyle differs from the pigeon's so that it is advantageous for bats to figure out potential navigation trajectories. For example, bats often reside in enclosed spaces like caves and knowing diverse navigation trajectories may be adaptive in such an environment. Bats are also known to be more specialized in maneuvering than flying compared to birds. Hence, navigation behaviors of bats may differ substantially from those of birds, suggesting that the simulation-selection function of the mammalian hippocampus has been preserved in bats during evolution.

Another possibility is a dissociation between the cognitive map and trajectory representations. Hippocampal replays of spatial trajectories during rest and sleep are yet to be found in the bat. Hence, it is possible that the bat hippocampus represents cognitive maps of the external world, which is manifested by clear place cells, but does not perform simulation-selection. Finally, we cannot exclude the possibility that place cells of the bat's hippocampus may serve functions other than cognitive map or trajectory representation.

A transformation of an existing function to another happens commonly during evolution. For example, auditory ossicles (tiny bones in the middle ear) have evolved to serve an auditory function (amplifying sound signals) from a jawbone that serves a feeding function. The exact functions of place cells found in the bat hippocampus remain to be clarified.

TUFTED TITMOUSE, A FOOD-CACHING BIRD

In contrast to bats and whales, some birds need particularly superb spatial memory for survival. Food-caching birds, such as the Clark's nutcrackers mentioned at the beginning of the chapter, store food at thousands of places in the fall and retrieve it successfully in the winter. An issue related to retrieving cached food from multiple locations is the sequence of visiting the locations. This is a well-known problem in artificial intelligence: the traveling salesman's problem, where the number of possible visiting sequences increases exponentially as the number of locations increases. It is then conceivable that evolutionary pressure might have implemented neural machinery to compute optimal visiting sequences in food-caching birds. In this regard, a study on the marsh tit, another food-caching bird species, has shown that it retrieves cached food in the sequence it is stored rather than randomly.[8] This finding suggests that the marsh tit retrieves cached food by considering not only stored locations but also navigation sequences. How do the birds solve these problems?

Even though we do not yet know, a recent study provides a clue to this issue. A team led by Dmitriy Aronov at Columbia University compared hippocampal neural activity between two bird species: food-caching and non-food-caching. Hippocampal neurons in the food-caching tufted titmouse showed place-specific activity, which is comparable to place-cell activity in the rat hippocampus. In contrast, the zebra finch, a species that does not store food, had much weaker place-specific hippocampal neuron activity.[9] An independent study also failed to find place cells in the quail's hippocampus (quails can fly even though they prefer to walk on the ground).[10] These findings suggest that the titmouse's hippocampus may have evolved to precisely represent the layout of the external space to meet the ecological demand (food caching) in a way similar to the spatial representation in the mammalian hippocampus. Considering that birds and mammals are separated by 310 million years of evolution, and that clear place cells are seldom found in non-food-caching birds (the pigeon, zebra finch, and quail), this is more likely to be an example of convergent evolution (wherein similar traits evolve independently) rather than a deeply preserved evolutionary trait. In other words, the titmouse's hippocampus has probably evolved to show place-specific activity independent of the evolution of the mammalian hippocampus.

Notably, for the first time in birds, Aronov's team detected sharp-wave ripples in both species (the tufted titmouse and zebra finch) during deep sleep. What are the functions of sharp-wave ripples in birds? Would the place cells of the titmouse's hippocampus fire along spatial trajectories during sleep like mammalian place cells, potentially aiding the bird in retrieving cached food in optimal visiting sequences? Then how would hippocampal neurons of the zebra finch fire during sharp-wave ripples? We currently have no answers to these questions. We cannot exclude the possibility that the titmouse's hippocampus has evolved to perform simulation-selection during sleep by taking advantage of existing sharp-wave ripples. If so, this could be an example of the convergent evolution of a highly complicated mental process.

EVOLUTION AND LIFE DIVERSITY

To summarize, it is not entirely clear how the hippocampus of two evolutionarily distant animal groups, birds and mammals, has evolved to handle spatial information in the process of adapting to diverse ecological environments. Evolution is not one-way traffic—it promotes changes in whatever direction that helps a species better survive. A species may lose an existing function or acquire a new one in this process. The history of such ongoing processes represents the current diversity of life. There are about thirty-five phyla in the animal kingdom. We belong to the phylum Chordata, which consists of three subphyla. Vertebrata is one. Of the remaining two, Urochordata is phylogenetically closer to vertebrates. Thus, Urochordata is the closest neighbor to vertebrates in terms of evolutionary history.

Figure 7.4 shows typical Urochordata, marine tunicates (sea squirts). Surprisingly, they are sessile (fixed in one place) and appear closer to sea anemones, animals located way lower on the phylogenetic tree than vertebrates. They do show all the characteristics of Chordata (i.e., notochord, dorsal nerve cord, gill slits, and postnatal tail) during development but lose these characteristics and motility during development. The case of Urochordata illustrates how far and diversely biological traits can change during evolution.

FIGURE 7.4. Marine tunicates. Figure reproduced from Nick Hobgood, "Tunicates," Wikimedia Commons, updated April 20, 2006, https://commons.wikimedia.org/w/index.php?curid=4590928 (CC BY-SA 3.0).

Perhaps the simulation selection function for potential spatial trajectories may have been lost or acquired during evolution. So far, our studies on hippocampal spatial representation and replays are limited to a handful of animal species. The picture will become clearer as we expand our hippocampal research to additional animal species.

PART III
The Neural Foundation of Abstraction

Chapter Eight

ABSTRACT THINKING AND NEOCORTEX

In chapter 7, we entertained the idea that the mammalian hippocampus has evolved to simulate (i.e., imagine) novel spatial trajectories to meet the unique ecological demands of land navigation. Human imagination, of course, is not limited to exploring novel spatial trajectories. We may imagine a story of a doctor who fell in love with two women during the turmoil of the Russian Revolution, imagine a new solution for the Riemann hypothesis, imagine artistic expressions that capture the fundamental loneliness of existence, imagine ourselves watching a clock while traveling at the speed of the light, and so on. Our unlimited capacity for imagination using highly abstract concepts is perhaps the most important element of our mental capacity that makes us particularly innovative.

ABSTRACT THINKING

How have humans, *Homo innovaticus*, acquired this capability? I argued that the hippocampus of land-navigating mammals evolved to simulate diverse navigation trajectories. If so, imagination per se is not a mental faculty unique to humans. Also, considering that the hippocampus processes not only spatial but also nonspatial information even in rats, the content of animal imagination probably includes both components.[1] It is likely then that the capability to imagine future events, both spatial and nonspatial, is a function shared by most mammals. If so, our superb capacity for innovation is unlikely due to our unique ability to imagine future events. Rather, it is more likely due to our exceptional capacity for high-level abstraction. In other words, the superiority of human innovativeness is likely the result of combining high-level abstract thinking with the already existing imagination function.

Abstract thinking refers to the cognitive capacity to derive general concepts or principles from individual cases. It allows us to form and manipulate concepts that are not tied directly to concrete physical objects or events, such as love, imaginary numbers, democracy, and free will. And it is not a unique mental faculty of humans; abstract thinking is pervasive in the animal kingdom. Nature is not random. Many animals figure out regularities in their environment, represent them as general rules, and behave accordingly to maximize survival and reproduction. Many animals also understand abstract notions such as numbers, are capable of transitive inference (e.g., if A > B and B > C, then A > C), and show behaviors suggesting self-awareness and theory of mind (the ability to attribute mental states, such as beliefs and intents, to others; e.g., "I know what you are thinking"; see chapter 1).[2] Neurophysiological studies have also found neural activity related to categories, high-order relationships, behavioral rules, and social interactions in rats and monkeys.[3] A specific example of the neural representation of an abstract concept (number sense) will be shown in chapter 9.

If we consider how a neural network stores experiences as memories, we can admit that abstraction is inevitable even in animals. Researchers generally believe that neural networks store multiple memories in an overlapping and distributed manner (see fig 8.1 and also fig. 4.3). If a neural network experiences many similar stimuli (such as many cars), those synapses (i.e., neural connections) activated by common features

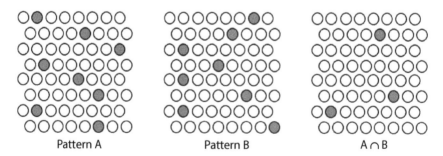

Pattern A   Pattern B   A ∩ B

FIGURE 8.1. Overlapping and distributed representation of memories in a group of neurons. One memory activates multiple neurons, and one neuron may participate in multiple memories.

of these stimuli (e.g., their "car-ness") will be strengthened more than other synapses because they will be repeatedly activated. Then those neurons having a sufficiently large number of the strengthened synapses will be preferentially activated by any of these stimuli. In other words, some neurons will be responsive to any car, be it familiar or novel, because they will be responsive to common features of cars after sufficient experiences. This enables us to recognize an object we have never seen before as a car. As this example illustrates, generalization, which is a form of abstraction, is thought to be a spontaneously emerging property of a neural network in animals and humans.

INNATE ABSTRACTION

The car example is a case for *empirical abstraction*. Neurophysiological studies have also found evidence for *innate abstraction* in animals. An eighteenth-century German philosopher, Immanuel Kant, claimed that humans are born with innate principles of cognition. This contrasts with the view of empiricism, which is traced back to Aristotle: knowledge comes primarily from sensory experiences rather than innate cognition. Kant, paralleling his proposal to the Copernican Revolution, claimed that we understand the external world based on not only experiences but also *a priori concepts*, things that can be understood independent of experience.[4] He argued that space and time are pure *intuitions* rather than properties of nature.[5] We consider space and time as physical entities that extend uniformly and infinitely in Cartesian coordinates because that's how we experience nature (i.e., the innate structure of cognition).

More than two centuries after Kant published the influential *Critique of Pure Reason* in 1781, neuroscientists found evidence for the intuition of spatial perception in the rat. The Norwegian neuroscientists Edvard Moser and May-Britt Moser, who shared a Nobel Prize with John O'Keefe in 2014, discovered "grid cells" in the rat entorhinal cortex, the main gateway of the hippocampus (the hippocampus communicates with the rest of the cerebral cortex mostly via the entorhinal cortex; see fig. 4.1).[6] A single grid cell fires periodically at multiple locations forming a hexagonal grid pattern of spiking activity as a rat moves around in each space (see fig. 8.2). This finding suggests a metric representation of the external space in the entorhinal cortex.

Importantly, a given grid cell maintains the same hexagonal grid firing pattern in all spaces. If you record the activity of one grid cell in two different

 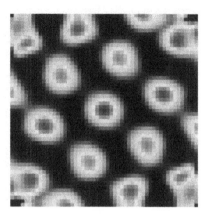

FIGURE 8.2. Activity of a sample grid cell. (Left) The black trace indicates a rat's movement trajectory, and gray circles indicate spike locations of the sample grid cell. Panel reproduced from Khardcastle, "Grid cell image V2," Wikimedia Commons, updated June 1, 2017, https://commons.wikimedia.org/wiki/File:Grid _cell_image_V2.jpg (CC BY-SA). (Right) A spatial autocorrelogram was constructed from the sample grid cell activity to reveal a periodic hexagonal firing pattern. Panel reproduced from Khardcastle, "Autocorrelation image," Wikimedia Commons, updated Jun 1, 2017, accessed Dec 21, 2022, https://commons.wikimedia.org /wiki/File:Autocorrelation_image.jpg (CC BY-SA).

rooms, the same pattern is observed in both (the generality of grid firing). The only difference is the orientation of the overall grid firing pattern across the two rooms (see fig. 8.3), indicating that the structure of spatial representation in the entorhinal cortex is identical regardless of where you are. Furthermore, if you record multiple grid cells simultaneously, their spatial

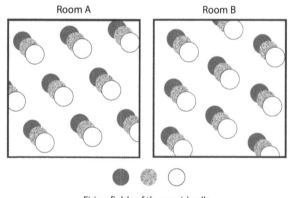

Firing fields of three grid cells

FIGURE 8.3. A schematic for grid firing patterns of three grid cells recorded in two different rooms.

relationships are preserved across multiple environments. In other words, all grid firing patterns rotate to the same degree when an animal is moved from one environment to another. This observation further corroborates the conclusion that the entorhinal cortex represents all external spaces in the same format, which is consistent with the notion of an a priori rule for spatial perception.

The entorhinal cortex consists of two divisions: medial and lateral. Grid cells are only found in the medial division. In the lateral division, neurons are responsive to objects encountered in each environment, such as a visual landmark, suggesting that they encode specifics of an environment.[7] The inputs from the lateral and medial entorhinal cortices converge in the hippocampus. This way, the hippocampus appears to combine two different types of inputs regarding spatial perception—one carries a general metric representation of space (medial entorhinal cortex) and the other carries information about specifics of an environment (lateral entorhinal cortex). This suggests that the hippocampus combines two different types of information: abstract structural knowledge and individual sensory experiences that are provided by the medial and lateral entorhinal cortex, respectively (see fig. 8.4). Studies in humans have shown that grid cells may encode not only spatial information but also conceptual knowledge in a gridlike activity pattern.[8] These results led Timothy Behrens and colleagues to propose that the entorhinal cortex's medial and lateral divisions, respectively, carry abstract structural knowledge and individual sensory experiences beyond the spatial domain.[9] This idea parallels Kant's proposal that we understand the world by the interplay between experiences and a priori concepts.[10]

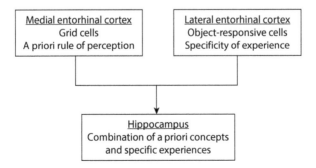

FIGURE 8.4. A schematic showing the information flow from the entorhinal cortex to the hippocampus.

HUMAN ABSTRACT THINKING

Humans, as well as other animals, are capable of abstract thinking. Researchers have learned a lot about the neural processes that underpin abstraction in animals. However, no other animal is capable of humans' high level of abstract thinking. Our capacity for high-level abstraction is overwhelming compared to other animals. Language is a prime example. We manipulate symbols according to grammatical rules in our daily use of language. Other animals, of course, can communicate with one another. No animal communication system, however, comes close to the complexity and flexibility of human language. In conclusion, animals are capable of abstract thinking and, most likely, imagination in the abstract domain. However, their abstract thinking and imagination would be much more limited than ours.

Our thinking process routinely crosses concrete and abstract domains. We are born with the ability to think abstractly, so we do not need to make an effort to do so. To illustrate this point, let's examine the "category mistake" proposed by the renowned English philosopher Gilbert Ryle.[11] Ryle introduced the idea in the process of refuting Cartesian dualism, which argues that the mind and the body are two separate entities. So what is the category mistake? Let's assume that your friend is showing you around her university. She shows you libraries, cafeterias, dormitories, laboratories, and lecture rooms. Now you ask, "But where is the university?" Of course, the university consists of all these things together. The university is a collection of institutions that may or may not be housed in their buildings. This example illustrates a semantic error in which things belonging to two different ontological categories (the university versus its buildings) are treated as though they belong to the same category. What is wrong with saying "there are three things in a field: two cows and a pair of cows"?[12] It's a category mistake, of course. Likewise, putting "mind" and "body" in the same category and assuming their existence as separate entities would be an example of a category mistake.

To me, these examples reveal our innate tendency to mix up concrete and abstract concepts and to consider abstract entities (such as "university" and "mind") as though they exist as concrete entities (such as "library" and "body"). This is probably a consequence of our ability to think seamlessly using both concrete and abstract concepts. Neural representations of concrete versus abstract entities may not be different according to the brain.

After all, those representations are merely implemented by changes in neural connections (see fig. 4.3) and patterns of neural activity. This perspective further promotes the view that human capacity for innovation is a natural consequence of adding the capacity for high-level abstraction, which is unique to humans, to the capacity for imagination, which is common to all mammals.

HOW MANY NEURONS IN THE CORTEX?

How have humans acquired the capacity for high-level abstraction? One characteristic of the human brain is its size. We have large brains compared to other animals including nonhuman primates. For instance, the human brain is three times larger than the chimpanzee's. A larger brain harnessing more neurons allows for higher processing power. In other words, size does matter for the relationship between the brain and intelligence.

Size isn't everything, however. Elephants and whales have larger brains and more neurons than we do. Nevertheless, our intelligence is superior to those animals. How is this possible? The comparative neuroanatomist Suzana Herculano-Houzel has spent years counting the numbers of neurons in different areas of the brain across a wide variety of animal species. Her studies indicate that the way neurons are distributed in the brain varies substantially across animal species. For example, elephants have a brain three times larger than ours and, accordingly, have more neurons in the brain than us. In the cerebellum, which plays a crucial role in movement and posture control, elephants have far more neurons (251 billion) than humans (69 billion).[13] The picture changes, however, if we examine the cerebral cortex. Humans have about three times as many neurons (16 billion) there as elephants (5.6 billion).[14] Humans, in fact, have more neurons in their cerebral cortex than any other animal on the planet. While the cerebellum in humans contains about 80 percent of the brain's neurons, it only constitutes 10 percent of the brain's mass. On the other hand, the cerebral cortex contains less than 20 percent of all neurons, yet it makes up approximately 80 percent of brain mass. This is due to the cerebral cortex having larger neurons, more neural connections, and a greater number of glial cells compared to the cerebellum.[15] As a result, the cerebral cortex has incredibly complex neural circuits. Considering the central role of the cerebral cortex in supporting advanced cognitive functions, it appears that humans are particularly intelligent and cognitively advanced as a result of their well-developed cerebral cortex.

Herculano-Houzel's studies indicate two advantages that helped humans to possess the largest number of neurons in the cerebral cortex.[16] First, we are primates. The packing density of neurons in the cerebral cortex is much higher in primates than in other mammals. Chimpanzees, much smaller in body size than elephants, have about the same number of cortical neurons (6 billion compared to 5.6 billion). Orangutans and gorillas, also much smaller than elephants, have more cortical neurons (9 billion). Thus, primates appear to have found an efficient way to pack far more neurons into the cerebral cortex than other animal groups during evolution.

Second, we have the largest brain of all primates. Combined with those highly packed cortical neurons, our big brain allows us to have the largest number of neurons in the cerebral cortex (16 billion) of any animal on earth (see fig. 8.5).

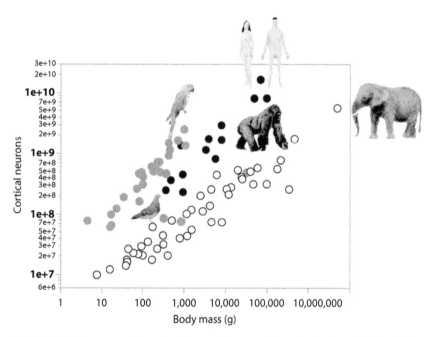

FIGURE 8.5. Body mass versus the number of neurons in the mammalian cerebral cortex or bird telencephalon. The scaling rule for the power function relating body mass to the number of cortical neurons varies across animal groups. Black, open, and gray circles indicate primate, nonprimate mammalian, and bird species, respectively. Figure adapted with permission from Suzana Herculano-Houzel, "Life History Changes Accompany Increased Numbers of Cortical Neurons: A New Framework for Understanding Human Brain Evolution," *Progress in Brain Research* 250 (2019): 183.

EXPANSION OF THE NEOCORTEX

The cerebral cortex can be categorized into distinct evolutionary divisions. These include the archicortex (hippocampus) and paleocortex (olfactory cortex), which represent older divisions, as well as the neocortex, which represents a newer division (note that *cerebral cortex* is commonly used to refer to the neocortex, as in chapter 4). Do we imagine freely using high-level abstract concepts because our hippocampus (archicortex) has special features that are absent in other animals? As we examined in chapter 7, the structure of the hippocampus is remarkably similar in all mammals including humans (see fig. 7.1). Also, as we discussed in chapter 1, damage to the hippocampus leaves most brain functions other than memory and imagination largely intact. Furthermore, hippocampal injuries change the content of mind-wandering from vivid and episodic to semantic and abstract.[17] Thus, it is not very likely that an idiosyncrasy of the human hippocampus is behind the unique human capacity for imagination using high-level abstract concepts.

One of the most outstanding differences between human and other animal brains is the relative size of the neocortex. The surface area of the hippocampus comprises 30–40 percent of the entire cortical surface area in rats but only 1 percent in humans. This is because the neocortex has expanded greatly in primates, especially humans. Primates evolved over 60 million years ago; apes evolved from African monkeys over 30 million years ago; great apes diverged from gibbons over 20 million years ago; humans (hominins) and chimpanzees split from the last common ancestors 6–8 million years ago.[18] The brains of our ancestors (hominins) did not grow much for a few million years after the split from chimpanzees. However, the hominin brain expanded dramatically, from about 350 grams to 1,300–1,400 grams, during the last two to three million years. This represents unusually rapid growth in the geological time scale. Three million years is only 5 percent of the history of primate evolution, but human brains quadrupled in size during this relatively short period.[19] The growth is largely due to the expansion of the neocortex, which takes up about 80 percent of the human brain by volume. Early mammals had little neocortex and about twenty distinct cortical areas, but present-day humans have about two hundred.[20] This suggests that the rapid expansion of the neocortex was the main driving force for human cognitive specialization.

84

THE NEURAL FOUNDATION OF ABSTRACTION

Even though the size of the neocortex varies widely across mammals, the basic circuit structure is largely similar across all areas of the mammalian neocortex (see fig. 8.6). This common neocortical circuit supports a plethora of brain functions such as seeing, listening, touch sensation, movement control, language comprehension, and reasoning. This suggests that the common circuit structure of the neocortex is an efficient, general-purpose processing module that can be used flexibly for a variety of purposes. In this regard, Bradley Schlaggar and Dennis O'Leary have shown that a strap of visual cortex (necessary for seeing) transplanted to the somatosensory cortex (necessary for touch sensation) in neonatal rats develops into a functional somatosensory cortex.[21] Thus, although there remains a possibility the human neocortex has unique features, such as specialized neurons, that contribute to our advanced cognitive abilities, it is more likely that human

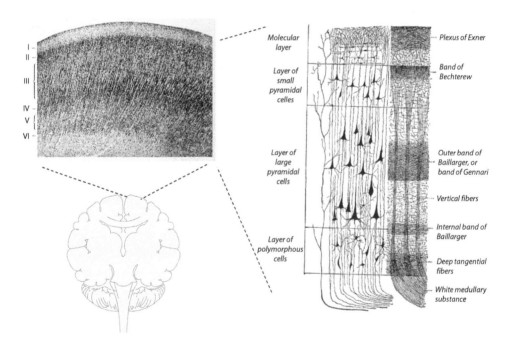

FIGURE 8.6. The six-layer organization of the neocortex. The left top panel is reproduced from Korbinian Broadmann, "File:Radial organization of tectogenetic layers in cerebral cortex, human fetus 8 months (K. Brodmann, 1909, p. 24, fig. 3).jpg," Wikimedia Commons, updated January 23, 2020, https://commons.wikimedia.org/wiki/File:Radial_organization_of_tectogenetic_layers_in_cerebral_cortex,_human_fetus_8_months_(K._Brodmann,_1909,_p._24,_Fig._3).jpg; the right panel is reproduced from Henry V. Carter, "File:Gray754.png," Wikimedia Commons, updated January 23, 2007, https://commons.wikimedia.org/wiki/File:Gray754.png.

neocortical evolution is driven by the duplication of existing circuits rather than the invention of new ones. Some genes regulate the activity of others. Some turn other genes on or off at specific times during brain development. Mutations in these genes could lead to dramatic changes in overall brain organization. Perhaps the evolutionary pressure during the last three million years favored the selection of mutations that expanded the neocortex by promoting the duplication of the common neocortical circuit.

We should keep in mind, however, that a large brain comes with a cost. It gives you more processing power, but more energy is needed for its maintenance, which is problematic considering the brain is a metabolically expensive organ. According to an influential theory, the expensive tissue hypothesis, humans solved this problem by consuming a high-quality diet and thereby reducing the size of another energy-demanding organ, the gut.[22] Many scientists agree that a high-quality diet or increased foraging efficiency probably helped the brain grow larger but disagree on how exactly this was achieved. According to an appealing hypothesis, the use of fire (i.e., cooking) increased the available caloric content in the food.[23] However, the brain grew larger steadily in the hominin lineage over the last three million years or so, whereas the earliest available evidence for the use of fire dates back only one million years. Also, it is unclear whether cooking dramatically enhances the caloric availability of meat.[24] Thus, the exact driving force for the expansion of the neocortex in the hominin lineage remains to be clarified.

HIPPOCAMPUS-NEOCORTEX DIALOGUE

The hippocampus and neocortex do not work independently but interact closely. The hippocampus is among the highest-order association cortices located far away from primary sensory and motor cortices. It interacts with the outside world primarily via its interactions with the neocortex. Hence, the content of memory and imagination the hippocampus deals with must be dictated by the information the neocortex provides. It is then plausible that the expansion of the neocortex advanced cognitive capability so that humans can imagine in spatial and nonspatial domains, including those dealing with high-level abstract concepts.

The following is a summary of my thoughts about how humans have acquired the capability to imagine freely using high-level abstract concepts. The imagination function of the hippocampus, i.e., the simulation-selection function, has evolved in land-navigating mammals to meet the need to represent optimal navigation routes between two arbitrary points in an area. The content of imagination then expanded from spatial trajectories to general episodes as the neocortex expanded. In particular, the explosive expansion of the neocortex has allowed the human brain to think and imagine using high-level abstract concepts (symbolic thinking). This has led to cultural and technological innovations including the organization of big societies.[25] One highlight of brain evolution for abstract thinking would be its linguistic capacity. Language allows humans to communicate abstract ideas such as theories and feelings. Written languages, in particular, accelerated innovation by facilitating knowledge accumulation (see fig. 0.1). These consequences of high-level abstract thinking probably acted synergistically to accelerate innovations leading to human civilizations that are unprecedented in the history of life on earth.

To summarize, human innovation is probably an outcome of acquiring the capability for high-level abstraction due to the expansion of the neocortex. Adding this capability to the already existing imagination function, which is common to all mammals, allowed humans to be truly innovative. But which aspect of the neocortical expansion is crucial for high-level abstraction? This issue will be explored in the following chapters.

Chapter Nine

PREFRONTAL CORTEX

Assuming that the human capacity for high-level abstraction is because of the expansion of the neocortex, which aspects of the human neocortex are critical for that ability? Does it feature special types of neurons or special neural circuits? Is there a special region in the human neocortex that is absent in other animals? Or does the capacity for innovation emerge from its global organization? We do not have clear answers to these questions, even though higher brain functions are generally believed to be outcomes of emergent properties of neural circuits rather than properties of individual neurons. One major reason for our ignorance on this matter is the absence of a decent animal model. Studies using animals greatly advanced our understanding of the neural bases of memory and imagination. In fact, for most brain functions, our understanding of their underlying neural mechanisms is owed greatly to animal studies. Obviously, this cannot be the case for the brain functions that are unique to humans, such as higher-order abstract cognition. Nevertheless, we are not without clues on this issue. Here, we will examine a few of these as samplers rather than exhaustively going over all of them. We will examine three clues: one derived from neuroscientific research (this chapter), another derived from paleoanthropological work (chapter 10), and the final one derived from neural network modeling (chapter 11).

THE NEURAL FOUNDATION OF ABSTRACTION

EXECUTIVE FUNCTION

Which brain structure would a neuroscientist say is most likely to play a crucial role in high-level abstract cognition in humans? Most neuroscientists would be reluctant to bet on one brain structure. Nevertheless, if they were forced to do so, perhaps a considerable number would pick the prefrontal cortex—the foremost part of the cerebral cortex (see fig. 9.1). Adjacent to the prefrontal cortex are premotor and supplementary motor cortices, which are known to play crucial roles in, roughly speaking, movement planning. These cortices project heavily to the primary motor cortex, which oversees the direct control of voluntary movement. As one may guess from this organization, the prefrontal cortex is an important brain structure for the control of behavior. However, the prefrontal cortex is not directly involved in the control of specific movement per se. Rather, the prefrontal cortex controls behavior flexibly according to internal goals and behavioral contexts.

The prefrontal cortex may promote a specific behavior in one circumstance but inhibit it in another. For example, in New York, you should look left before crossing a street to check for passing vehicles. However, in London, you should look right before stepping off the curb. Hence, in London, New Yorkers must suppress their habitual behavior of looking left to check for cars before crossing the street. This simple example illustrates the importance of a context-dependent control of behavior in our daily lives.

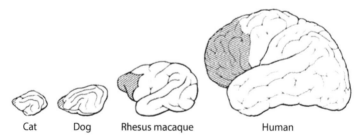

Cat   Dog   Rhesus macaque   Human

FIGURE 9.1. Prefrontal cortex, indicated in shading, in several different animal species. Figure reproduced with permission from Bruno Dubuc, "The Evolutionary Layers of the Human Brain," The Brain from Top to Bottom, accessed December 11, 2022, https://thebrain.mcgill.ca/flash/a/a_05/a_05_cr/a_05_cr_her/a_05_cr_her.html.

Controlling behavior flexibly according to an internal goal in an ever-changing world is not a simple job. A diverse array of functions is required to be able to do that. Not surprisingly, previous studies have found that the prefrontal cortex serves a wide range of cognitive functions: working memory, reasoning, planning, decision-making, and emotional control; the term *executive functions* collectively describes the functions of the prefrontal cortex.

It controls and coordinates other functions of the brain so that our behaviors, or suppression of them, serve the role of achieving our internal goals. It would be impossible without the prefrontal cortex to work day and night to achieve your long-term goal, such as getting a doctoral degree in neuroscience in five years. As you may have guessed, the prefrontal cortex is well developed in primates, especially in humans.[1] One could argue that the prefrontal cortex differentiates humans from other animals. When we look at the behaviors of people whose prefrontal cortex has been damaged in some way, this becomes abundantly clear.

PHINEAS GAGE

Phineas Gage was a foreman in a crew cutting a railroad bed in 1848 in Vermont. He was twenty-five years old at the time. The crew blasted rocks by boring a hole deep into a rock, adding explosive powder, and then packing inert material, such as sand, into the hole with a tamping iron. On September 13, probably because Gage inadvertently omitted adding sand, the powder exploded because of a spark between the tamping iron and the rock. The tamping iron, just over three and a half feet long, passed through Gage's head and landed over eighty feet away. Miraculously, he survived the accident and, surprisingly, could speak and walk a few minutes later. So what were the consequences of this accident? Gage had no problem seeing, listening, speaking, or moving around. The tamping iron damaged the foremost region of the brain, the prefrontal cortex (see fig. 9.2), indicating that this region of the brain is not needed for brain functions such as life support, sensation, motor control, or language.

Nevertheless, Gage's personality changed dramatically following the accident. He used to be a smart, energetic, and shrewd young man with a promising future and was considered the most efficient and capable man in the company. However, after the accident, he was "no longer Gage" to

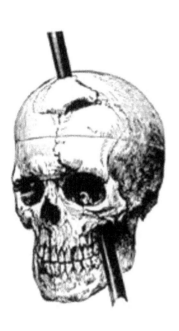

FIGURE 9.2. Brain injury of Phineas Gage. (Left) A tamping iron traveled through Gage's brain. Panel reproduced from John M. Harlow, "File:Phineas gage—1868 skull diagram.jpg," Wikimedia Commons, updated October 25, 2007, https://commons.wikimedia.org/wiki/File:Phineas_gage_-_1868_skull_diagram .jpg; (right) Gage holding the tamping iron that injured him. Panel reproduced from "File:Phineas Gage GageMillerPhoto2010-02-17 Unretouched Color Cropped.jpg," Wikimedia Commons, updated August 2, 2014, https://commons.wikimedia.org/wiki/File:Phineas_Gage_GageMillerPhoto2010-02-17_Unretouched _Color_Cropped.jpg.

his friends. Dr. John Harlow, who treated Gage for a few months, recorded the following notes:

> The equilibrium or balance, so to speak, between his intellectual faculties and animal propensities, seems to have been destroyed. He is fitful, irreverent, indulging at times in the grossest profanity which was not previously his custom, manifesting but little deference for his fellows, impatient of restraint or advice when it conflicts with his desires, at times pertinaciously obstinate, yet capricious and vacillating, devising many plans of future operation, which are no sooner arranged than they are abandoned. . . . A child in his intellectual capacity and manifestations, he has the animal passions of a strong man.

Previous to his injury, though untrained in the schools, he possessed a well-balanced mind, and was looked upon by those who knew him as a shrewd, smart business man, very energetic and persistent in executing all his plans of operation. In this regard his mind was radically changed, so decidedly that his friends and acquaintances said he was "no longer Gage."[2]

PERSISTENCE AND FLEXIBILITY

To achieve a long-term goal, one must be persistent yet still flexible rather than "obstinate, yet capricious," which was how Harlow described Gage. Suppose you need to buy groceries to make dinner for a friend. Your plan is to buy steak and garlic because your friend's favorite dish is garlic steak. You arrive at the store and find many delicious-looking food—bread, pasta, fish, and fruit—on your way to the meat section. You must be persistent and ignore all these distracting stimuli and achieve your original goal (getting the ingredients for the garlic steak). This wouldn't be difficult for neurotypical people, but it would be for those living with prefrontal cortex damage since they are easily distracted by intervening stimuli. In this example, they may buy any items that catch their attention, such as salmon and pasta, instead of steak and garlic. In fact, some may not even be able to get to the store in the first place. They may get on a random bus that happens to stop by them on their way. This example illustrates the importance of the prefrontal cortex in making us persist in achieving a long-term goal.

The prefrontal cortex is important not only for persistence but also for flexibility. If a certain behavior is no longer effective in achieving your current goal, you must switch to another type. If you stick to behavior that is no longer effective, you would be considered obstinate rather than persistent. People with prefrontal cortex damage have trouble changing their behavior according to changes in the environment. This can be tested with the Wisconsin card sorting test, a neuropsychological test frequently used to measure perseverative behaviors (insistence on wrong behavior). In this task, the subject must classify a stimulus card according to a hidden rule that changes over time. Figure 9.3 shows a sample computer screen of the task. The subject must match the sample card shown in the bottom right corner of the screen (the card with two green squares) to one of the four cards on top. Possible hidden rules are the color, quantity, and shape of the figures in the card. The hidden rule is not revealed to the subject. The subject is only told

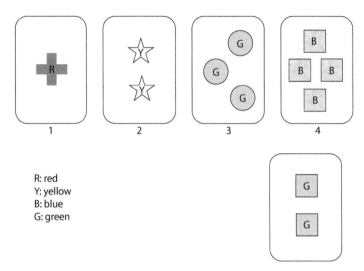

R: red
Y: yellow
B: blue
G: green

FIGURE 9.3. Wisconsin card-sorting test.

whether their match was right or wrong. In the example shown, the right match may be the second card (hidden rule: number), the third card (hidden rule: color), or the fourth one (hidden rule: shape). People with no prefrontal cortex damage quickly adjust their matching strategy based on the feedback ("wrong") when the current hidden rule changes. However, after getting that feedback (i.e., when the hidden rule changes), people with prefrontal cortex damage tend to stick to the old matching strategy that no longer yields a correct match.

PREFRONTAL CORTEX AND ABSTRACTION

Abstraction is essential for guiding behavior according to a long-term internal goal, for changing behavior flexibly according to hidden rules, and for reasoning and planning, all of which are functions of the prefrontal cortex. Indeed, patients with prefrontal cortex damage show impaired abstraction in various clinical tests.[3] For example, in a proverb interpretation task, patients may have trouble understanding a statement in an abstract sense. "Rome was not built in a day" may be interpreted in a concrete sense (e.g., the time it takes to complete the infrastructure of the Roman Empire) rather than an abstract sense (the time it takes to accomplish great achievements).[4]

Consistent with these neuropsychological findings, physiological studies of the prefrontal cortex have found abstraction-related neuronal activity even in animals. Neurophysiologists who study the prefrontal cortex joke that the prefrontal cortex is like a department store, meaning that you can usually find whatever neural signals you are looking for if you stick a microelectrode in the prefrontal cortex. Why is that so? As we saw with Phineas Gage, the prefrontal cortex is not specialized to process specific sensory or motor signals. Rather, it represents the information necessary to complete a given task successfully. The information varies widely depending on the current goal and includes not only specific sensorimotor signals but also abstract concepts, such as task rules, that are needed to achieve the goal. In other words, the prefrontal cortex appears to be a general-purpose controller that flexibly and selectively represents the information essential to control a goal-directed behavior. Hence, if you train an animal to perform a certain behavioral task, you are likely to find concrete as well as abstract neural signals essential to perform the task in the prefrontal cortex.

Numerous studies have shown diverse types of abstraction-related neural activity in the prefrontal cortex of animals. Here, as an example, let's examine the representation of number sense in the monkey's prefrontal cortex. Numerosity, the number of elements in a set, is an abstract category that is independent of physical characteristics of elements. For example, "three-ness" is an abstract concept that is applicable to any set containing three elements regardless of their modality. Neurophysiological studies in monkeys have found neurons in the prefrontal cortex that are selectively tuned to the number of items. In one study, Andreas Nieder trained rhesus monkeys to discriminate the number of sequentially presented items to obtain a reward. The items were presented in two different formats: sound pulses (auditory) or black dots (visual). He found neurons in the prefrontal cortex that are tuned to the number of items irrespective of the modality of the stimuli. Each numerosity-coding neuron is preferentially responsive to a particular number so that a group of prefrontal cortex neurons can provide unambiguous information about the number of successive stimuli presented to the monkey.[5] This and other related studies clearly show that the monkey's prefrontal cortex represents number sense, which is an abstract category beyond physical characteristics of elements.[6]

Neural activity related to abstract concepts, such as task rule, task structure, and behavioral context, has been found even in the rat and mouse prefrontal cortex.[7] A rodent's prefrontal cortex is tiny compared to a human's.

Not only the absolute size but also the size relative to the whole brain is much smaller for a rodent than a human. Even so, rodents do show considerable levels of sophisticated behavioral control that are needed to achieve a long-term goal. For example, damages in the prefrontal cortex impair flexibility in a rodent version of the Wisconsin card-sorting test.[8] These animal studies show clearly that the rodent prefrontal cortex, very underdeveloped compared to the human prefrontal cortex, supports flexible control of behavior and represents abstract concepts. It is conceivable then that the well-developed human prefrontal cortex may represent much higher-level abstract concepts than those found in animals.

One domain of behavioral control where high-level abstraction is necessary would be human social behavior. Gage's behavior after his tragic accident was far from expected for a normal adult. It would be extremely difficult for "a child in his intellectual capacity" to behave properly in a complex human society. Our social behaviors are outcomes of careful considerations of subtle social conventions that vary widely according to social contexts; therefore, abstract thinking is essential. For example, we routinely use high-level abstract notions such as public order, rudeness, and social reputation to decide how to behave every day. It may be that the human capacity for high-level abstract thinking owes a lot to, among other things, the evolutionary pressure to form and represent high-level abstract concepts for the proper control of behavior in a complex human society.

To summarize, there are good reasons to suspect that the well-developed human prefrontal cortex is a core region of the neural system forming and representing high-level abstract concepts. Of course, it is not likely that the expansion of the prefrontal cortex is solely responsible for our enhanced abstract thinking capacity. Other changes in the brain are likely to be crucial. For example, the parietal cortex is activated together with the prefrontal cortex during various abstraction tasks.[9] Thus, understanding how the prefrontal cortex interacts with other brain areas is also important to understand the neural basis of human abstraction.

FRONTOPOLAR CORTEX

Another important issue is which subregion of the prefrontal cortex plays the most crucial role in abstraction. The prefrontal cortex is a large area consisting of multiple subregions serving distinct functions, and it is unclear how the abstraction function is distributed across these subregions.

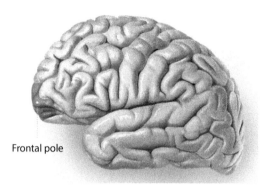

Frontal pole

FIGURE 9.4. Frontopolar cortex. Figure reproduced from Frank Gaillard, "Frontal Pole," Radiopaedia.org. rID: 46670, accessed April 6, 2023, https://radiopaedia.org/articles/34746 (CC-BY-NC-SA).

Brain imaging studies suggest that more anterior (frontal) parts of the prefrontal cortex serve more abstract thinking and more posterior (rear) parts serve more concrete thinking.[10] If so, future studies focusing on the frontopolar cortex (see fig. 9.4), the most anterior region of the prefrontal cortex, may provide important clues to our quest to elucidate the neural basis of human innovation.

Chapter Ten

THE HUMAN REVOLUTION AND ASSOCIATED BRAIN CHANGES

HUMAN REVOLUTION

Looking back at the past may help us guess what critical changes in the brain enabled high-level abstract thinking in *Homo sapiens*. Specifically, comparing fossilized skulls and archaeological evidence may provide a clue on the neural basis of high-level abstraction. In this regard, archaeological studies have identified a suite of innovations that characterize modern human behavior, such as verbal communication, exchange of goods, sophisticated burials, artistic expression, and organized societies, in Eurasia during the Upper Paleolithic, about forty to fifty thousand years ago. This was once known as the "human revolution" because of a relatively sudden appearance of a *package* of modern human behavior, even though archeologists have found earlier sporadic evidence for some of these characteristics.[1]

These archaeological findings strongly suggest that our Eurasian Upper Paleolithic ancestors had a cognitive capacity not inferior to that of modern humans. Figure 10.1 shows some of the Upper Paleolithic paintings discovered in Europe. Many tools, sculptures, and wall paintings from this era are artistically excellent even by today's standards. Symbolic and abstract thinking are among the core cognitive functions associated with art. For example, the Venus figurine (see fig. 10.2), commonly found in Upper Paleolithic sites throughout Eurasia, is thought to symbolize fertility. One popular theme

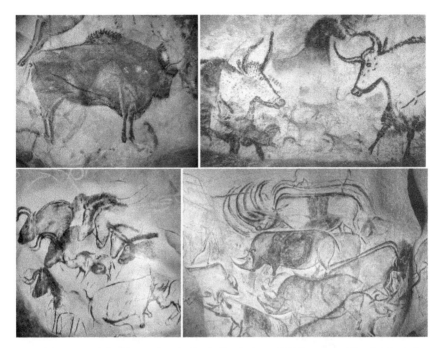

FIGURE 10.1. Upper Paleolithic paintings (replicas): Altamira (upper left), Rameessos, "Cave Painting in the Altamira Cave," World History Encyclopedia, January 6, 2015, https://www.worldhistory.org/image/3537/cave-painting-in-the-altamira-cave/; Lascaux (upper right), Prof saxx, "Cave Painting in Lascaux," World History Encyclopedia, January 7, 2015, https://www.worldhistory.org/image/3539/cave-painting-in-lascaux/ (CC BY-NC-SA); and Chauvet Cave (lower left), T. Thomas, "Cave Paintings in the Chauvet Cave," World History Encyclopedia, July 16, 2014, https://www.worldhistory.org/image/2800/cave-paintings-in-the-chauvet-cave/ (CC BY-NC-SA) and (lower right) Patilpv25, "Panel of the Rhinos, Chauvet Cave (Replica)," World History Encyclopedia, February 10, 2017, https://www.worldhistory.org/image/6350/panel-of-the-rhinos-chauvet-cave-replica/ (CC BY-SA).

of art throughout the history of mankind is sex and fertility, i.e., the preservation of our species. The Venus figurine may be an ancient version of present-day African fertility dolls, which symbolize a fertility goddess.[2] In general, an artist needs the cognitive capacity to project their symbolic content onto a mental image while alternating between their own and potential viewers' perspectives in order to communicate ideas, feelings, and experiences.[3] The Upper Paleolithic art hints at our ancestors' superb symbolic and abstract-thinking abilities.

Ornament burial is another line of evidence for the behavioral modernity of our Upper Paleolithic ancestors. Graves are regarded as clear evidence of religious beliefs, a hallmark of modern human behavior. Figure 10.3 depicts a sophisticated ornamental burial from that period. Even though the view that

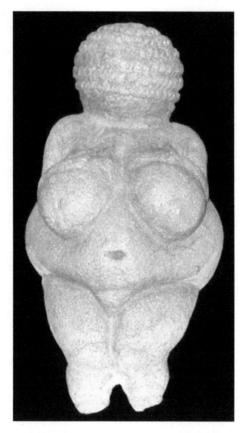

FIGURE 10.2. Venus of Willendorf. Figure reproduced from Oke, "Venus of Willendorf," Wikimedia Commons, updated October 29, 2006, https://en.wikipedia.org/wiki/Venus_figurine#/media/File:Wien_NHM_Venus _von_Willendorf.jpg (CC BY-SA).

burials became ubiquitous and elaborate only during the Upper Paleolithic appears to be an oversimplification, sophisticated and elaborate burials such as the one shown suggest that our Upper Paleolithic ancestors thought about what happens after death and believed in the transmigration of the soul as many modern humans do.[4] *Homo sapiens*' spiritual awareness and religiousness may not have changed much since then.

The archeological phenomena in the Upper Paleolithic indicate that our ancestors had well-developed cognitive capacity, including high-level symbolic thinking. If an infant of that period was teleported to the present day, he or she could grow up to be a scientist, artist, writer, or politician, just like any other modern infant. Compare the two drawings in Figure 10.4.

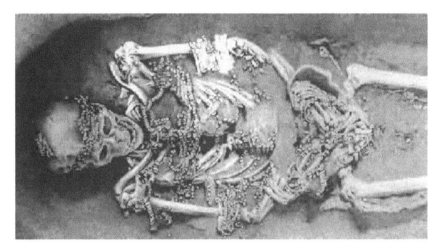

FIGURE 10.3. An Upper Paleolithic burial was found at Sungir, Russia. Figure adapted from "Red Haired Mummy," Mummipedia Wiki, updated June 8, 2013, https://mummipedia.fandom.com/wiki/Red_Haired _Mummy (CC BY SA).

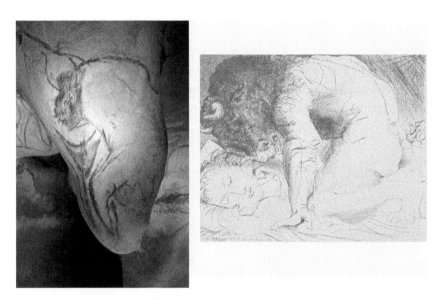

FIGURE 10.4. The beast within. (Left) Charcoal drawing found in Chauvet Cave, France. Panel reproduced from Claude Valette, "File:20 TrianglePubienAvecTêteDeBison&JambeHumaine.jpg," Wikipedia Commons, updated March 4, 2016, https://commons.wikimedia.org/wiki/File:20_TrianglePubienAvecT%C3%AAteDeB ison%26JambeHumaine.jpg (CC BY-SA). (Right) Pablo Picasso's *Minotaur carressant une dormeuse*. Drypoint print, 1933. Panel reproduced with permission (Society of Artist's Copyright of KOREA).

The left is a charcoal drawing found in the Chauvet Cave, France, which is estimated to be about thirty thousand years old. A bison-headed man is looming over a woman's naked body that has an emphasized pubic triangle. The right is a drawing by Pablo Picasso, *Minotaur carressant une dormouse*. The resemblance is striking. It appears that both represent a fantasy embedded deep in the male psyche ("the beast within"). The Chauvet Cave was not discovered until after Picasso's death. Thus, Picasso created the piece without being aware of that similar drawing. However, he visited the Lascaux caves in southwestern France and saw wall paintings like the ones shown in figure 10.1. Picasso, impressed by them, reportedly said to his guide, "They've invented everything."[5]

ANATOMICALLY MODERN HUMANS

It was generally believed until the late 1980s that modern *Homo sapiens* appeared forty to fifty thousand years ago in Eurasia (European model). This is when the Cro-Magnons appeared and Neanderthals disappeared. Cro-Magnons were tall and their brains were about 30 percent larger than those of modern humans. The skulls of Cro-Magnons and modern humans are round-shaped, while those of Neanderthals are much more elongated from front to back (see fig. 10.5). It is unclear why our ancestors survived while Neanderthals did not, but one possibility is that Cro-Magnons had a cognitive advantage.[6] Because their foreheads were flatter, and because the prefrontal cortex is a critical brain area for many highly cognitive functions,

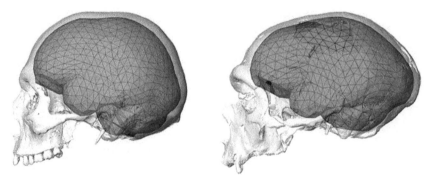

FIGURE 10.5. Differences in Neanderthal and modern human skulls. Figure reproduced from Simon Neubauer, Jean-Jacques Hublin, and Philipp Gunz, "The Evolution of Modern Human Brain Shape," *Science Advances* 4, no. 1 (2018): eaao5961 (CC BY-NC).

as discussed in chapter 9, it was natural to suspect that Cro-Magnons had advanced cognitive capacity due to their well-developed prefrontal cortex. According to this viewpoint, the human revolution coincides with the expansion of the prefrontal cortex.[7]

Subsequent discoveries indicate, however, that the story is not that simple. For example, the fossil records of anatomically modern *Homo sapiens* found in the Qafzeh and Skhul caves in Israel were initially thought to be about 40,000 years old according to the European model. However, a more precise estimation based on heat-induced emission of light (this method is called thermoluminescence dating: energized electrons accumulate over time in proportion to the radiation a specimen received from the environment) in the late 1980s indicated that the fossil records are about 100,000 years old.[8] Subsequent findings of new fossil records pushed back the origin of anatomically modern humans earlier and earlier. Archeological find-ings indicate that anatomically modern *Homo sapiens* appeared at least 200,000 years ago, and the fossil records for more primitive *Homo sapiens* found in Africa date back to about 300,000 years ago.[9] DNA studies also support the idea that anatomically modern humans appeared in Africa much earlier than the Upper Paleolithic.[10] It is now well agreed that anatomically modern *Homo sapiens* appeared in Africa long before the Upper Paleolithic and spread throughout Eurasia over thousands of years, replacing other early humans such as Neanderthals (the "Out-of-Africa" model).[11]

ORIGIN OF BEHAVIORAL MODERNITY

Thus, there appears to be a time lag between the appearance of anatomically modern humans and modern human behavior. What then was the main driving force for its sudden appearance? Some scholars advocate a neuro-logical change. A fortuitous mutation may have induced a subtle but critical neurological change to promote cognitive capability without a major change in the shape of the skull.[12] Alternatively, environmental and social factors, such as environmental stress and increased population density, rather than a biological change, might be the main driving force.[13]

Some researchers deny the concept of a human revolution, arguing that characteristics of modern human behavior appeared over a long period. They contend that technological and cultural innovations appeared and dis-appeared multiple times in the history of *Homo sapiens* due to factors such as population loss and environmental instability. This, along with the loss of archaeological records, may explain why there is only patchy evidence for behavioral modernity before the Upper Paleolithic.[14] It would be difficult to reconstruct events that occurred tens of thousands of years ago. We must wait until archaeologists and paleoanthropologists present compelling evidence supporting a particular hypothesis. Nonetheless, it appears that our ancestors acquired the capacity for high-level abstraction at least by the Upper Paleolithic.

PALEONEUROLOGY

Paleoneurology is the study of brain evolution by the analysis of the endocast, which is the internal cast of the skull. It would be difficult to detect a subtle neurological change during evolution by examining the endocast. However, major changes in brain structure are likely to occur alongside corresponding changes to it. Hence, if *Home sapiens'* cognitive capacity has reached the level of modern humans as a result of major structural changes in the brain, paleoneurological research may be able to provide valuable insights into them. In this regard, studies in paleoneurology have identified changes in the size and shape of the braincase during human evolution.[15]

Let's look at a recent study published in 2018. Three scientists at the Max Planck Institute for Evolutionary Anthropology in Germany—Simon Neubauer, Jean-Jacques Hublin, and Philipp Gunz—carefully analyzed twenty *Homo sapiens* fossils belonging to three different periods (300,000–200,000, 130,000–100,000, and 100,000–35,000 years ago). They found that the brain shape of only the third group, the most recent fossils, matches that of present-day humans.[16] This indicates that *Homo sapiens'* brain shape reached the modern form between 100,000 and 35,000 years ago. Because this period corresponds to the emergence of modern human behavior in the archaeological record, it is possible that these neurological changes played a significant role in the development of modern human behavior.

How exactly does the brain shape of the third group differ from the other two? The size of the brain was similar across all three groups, but the shape of the brain became more globular (see fig. 10.6). Further analysis indicated bulging of two brain regions, the parietal cortex and cerebellum (see fig. 10.7), during the brain globalization process.[17] Here we will focus on the parietal cortex because neocortical expansion is likely to be responsible for advanced cognitive capability in humans (see chapter 8), even though

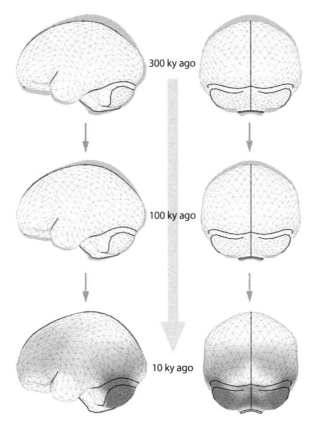

FIGURE 10.6. Brain globalization during evolution. Figure reproduced from Neubauer, Hublin, and Gunz, "The Evolution of Modern Human Brain Shape," eaao5961 (CC BY-NC).

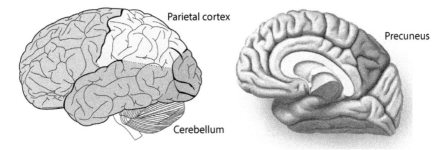

FIGURE 10.7. Parietal cortex, cerebellum, and precuneus. (Left) A side view of the brain showing the parietal cortex and cerebellum. Panel adapted from Henry V. Carter, "File:Lobes of the brain NL.svg," Wikidemia Commons, updated January 15, 2010, https://commons.wikimedia.org/wiki/File:Lobes_of_the_brain_NL.svg (public domain). (Right) A midline view of the brain showing the precuneus. Panel reproduced from Dayu Gai, Fabio Macori, and C. Worsley, "Precuneus," Radiopaedia, updated September 2, 2021, https://doi.org /10.53347/rID-38968 (CC BY-NC-SA).

we cannot reject the possibility that changes in the cerebellum contributed greatly to the appearance of modern human behavior.

Note that it is difficult to get a clear picture of the shape of the brain from the shape of the braincase.[18] Moreover, the parietal cortex is a wide area. Nevertheless, studies have shown that a particular region of the parietal cortex, namely the precuneus (see fig. 10.7), is proportionally much larger in humans than in chimpanzees;[19] its size is also correlated with the parietal bulging in modern humans.[20] Based on these findings, Emiliano Bruner at Centro Nacional de Investigación sobre la Evolución Humana (National Center for Research on Human Evolution) in Spain proposed that precuneus expansion may be associated with recent human cognitive specialization.[21]

PRECUNEUS

Unfortunately, we know relatively little about the precuneus. It is rare to find a human patient with selective damage to the precuneus (such as by stroke), and there are very few studies on this enigmatic brain region. Andrea Cavanna and Michael Trimble, in a paper published in 2006, reviewed published human brain imaging studies and grouped four different mental processes the precuneus appears to be involved in: visuospatial imagery (spatially guided behavior), episodic memory retrieval (remembering autobiographical memory), self-processing (representation of the self in relationship with the outside world), and consciousness.[22] Thus, the precuneus is activated in association with a diverse array of cognitive processes, which is not surprising given that it is an association cortex far removed from primary sensory and motor areas and extensively connected with higher association cortical areas.

Subsequent studies additionally found the involvement of the precuneus in artistic performance and creativity. For example, the gray matter[23] density of the precuneus is higher in artistic students than in non-artistic students,[24] and the functional connectivity (the correlation in activity over time between two brain regions, which is an index of how closely two brain areas work together) between the precuneus and thalamus is stronger in musicians than in nonmusicians.[25] Also, the activity of the precuneus is correlated with verbal creativity[26] and ideational originality.[27] These findings suggest that the expansion of the precuneus may have been a significant stage in the evolution of the human capacity to form high-level abstract concepts.

As an alternative possibility, the human precuneus may facilitate the use, rather than the formation, of high-level abstract concepts. The precuneus is a "hot spot" of the default mode network. It consumes more energy (glucose) than any other cortical area during conscious rest.[28] As we examined in chapter 1, the default mode network is activated when we engage in internal mentation (conceptual processing) and deactivated when we pay attention to external stimuli (perceptual processing). Also notable is that the precuneus is activated across multiple behavioral states. According to brain imaging studies, the human brain consists of multiple, overlapping, and interacting neural networks that appear to serve distinct functions. One of them is the default mode network. Other well-characterized task-associated networks include the central executive network (or frontoparietal network; its major components are the dorsolateral prefrontal and posterior parietal cortices), which is active in cognitively and emotionally challenging situations.

The default mode network and task-associated networks were initially thought to be segregated and antagonistic. However, later studies have shown that a given brain area may be activated during both quiet rest and task performance, indicating that the two networks can overlap. Amanda Utevsky, David Smith, and Scott Huettel examined the brain activity of about two hundred human subjects while performing three reward-based decision tasks as well as during quiet rest in a 2014 study.[29] They identified one brain region that showed systematic state-dependent functional connectivity: the precuneus. It was the only brain area that increased functional connectivity with the central executive network during task performance and with the default mode network during rest. This suggests that the precuneus may play an important role in coordinating activity between various large-scale networks. Given that creative cognition involves dynamic interactions between large-scale brain networks including the default mode and central executive networks,[30] the precuneus may play a pivotal role in generating novel and innovative ideas.

Collectively, these neuroscientific findings raise the possibility that the human precuneus may facilitate the *use* of high-level abstract concepts during imagination by coordinating the activity of multiple large-scale networks and linking distinct subnetworks of the default mode network. It may act as an efficient linking channel between two brain systems critical for prompting imagination (such as the hippocampus) and handling high-level abstract concepts (such as the prefrontal cortex). Currently, the potential contributions of the precuneus to the human capacity for innovation we discussed so far are only speculative possibilities. Further studies are needed to reveal whether the expansion of the precuneus was a critical step toward the "great leap forward" of the cognitive capacity of modern humans.[31]

Chapter Eleven

DEEP NEURAL NETWORK

In chapters 9 and 10, we examined two different brain regions, the prefrontal cortex and the precuneus, whose expansions may have been pivotal for the unique human capacity for high-level abstraction. Here, we will entertain the possibility that this human capacity is owed primarily to the overall organization of the brain rather than the expansion of a particular brain region. For this, let's turn our attention to artificial neural networks. As a major branch of artificial intelligence, they perform computations based on architectures that mimic biological neural networks. Historically, there have been four waves of artificial neural network research since the 1940s. The third one was in the 1980s. I began my neuroscience career as a graduate student in California around this time. I still vividly remember the excitement they generated at the time.

It was, and still is, difficult to understand the brain because of its complexity. The human brain has approximately a hundred billion neurons and a hundred trillion connections, and many different types of neurons form a vastly complex system of connections. It is therefore a daunting challenge to unravel the way this complicated system processes information to guide behavior in an ever-changing world. Experimental neuroscientists study the brain directly, many using animal models that have somewhat less complex brains. For example, neurophysiologists place microelectrodes in the brain to monitor activity of single neurons while an animal is engaged in certain behaviors, such as deciding which of two available levers to press. As an alternative approach to understand the brain, computational neuroscientists study what artificial neural networks can do and how they do the job.

In the 1980s, neuroscientists and engineers alike hoped that artificial neural networks would lead to significant advances in our understanding of the brain and offer novel solutions to challenging AI problems such as visual object classification. To their disappointment, the initial excitement gradually subsided because, despite their significant contributions, artificial neural networks did not live up to expectations. The fourth wave of artificial neural networks is currently in progress, and "deep learning" is driving this wave.

DEEP LEARNING

Artificial neural networks are made up of layers of units (or nodes or artificial neurons) that behave according to simplified rules derived from biological neurons (such as "integrate and fire," in which each unit sums active inputs and emits outputs when a threshold is crossed). Typical artificial neural networks have an input layer, an output layer, and a variable number of hidden layers (see fig. 11.1). So, what exactly is deep learning in neural networks? It refers to learning performed by a multi-hidden-layer artificial

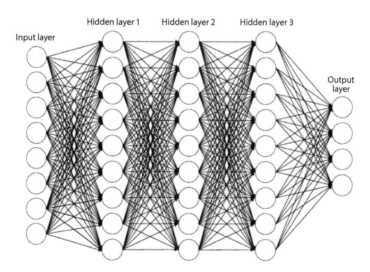

FIGURE 11.1. A schematic for an artificial neural network. Figure reproduced from Massimo Merenda, Carlo Porcaro, and Demetrio Iero, "Edge Machine Learning for AI-Enabled IoT Devices: A Review," *Sensors (Basel)* 20, no. 9 (April 2020): 2533 (CC BY).

neural network. Hidden layers enable a neural network to perform complex, nonlinear computations. The depth of a network refers to the number of hidden layers.

In the early 2010s, scientists found that performance of artificial neural networks in certain applications could be enhanced dramatically by adding more hidden layers (i.e., making the networks deeper). Deep neural networks started outperforming other methods in prominent image analysis benchmarks, most notably the ImageNet Large Scale Visual Recognition Challenge in 2012, and these achievements sparked the artificial intelligence "deep learning revolution."[1] Deep learning is now used in almost every field of AI research.

How do the hidden layers function? Multiple hidden layers allow for multiple levels of abstraction, which is useful for hierarchical feature extraction. Assume you want to train a neural network to recognize animal species from their photographs (see fig. 11.2). Once you train it with some sample animal

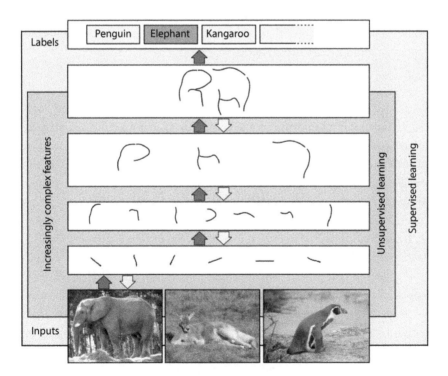

FIGURE 11.2. Learning feature hierarchy by a deep neural network. Figure reproduced with permission from Hannes Schulz and Sven Behnke, "Deep Learning: Layer-Wise Learning of Feature Hierarchies," *Künstliche Intelligenz* 26, no. 4 (2012): 357.

images (elephants, seals, etc.), then you test it with new images of these animals. A shallow neural network (e.g., one hidden layer) can be trained to perform this task, but the number of required feature detectors in the hidden layer grows exponentially with the number of inputs. A deep neural network can alleviate this problem.

In the example shown in fig. 11.2, a six-layer feed-forward neural network successfully classifies animal species from new images by forming a feature hierarchy.[2] Early hidden layers represent low-level features such as localized edges, which are progressively combined to represent more abstract features in late hidden layers, eventually representing high-level features for animal species prototypes (e.g., "elephant-ness"). This is a powerful and efficient way of implementing visual object recognition. Note that no information about animal species was provided to the hidden layers during training. The neural network organizes itself to hierarchically represent features to achieve the task. In fact, this line of neural network research was inspired by findings from visual neuroscience. Neurons in the early stages of visual processing are responsive to low-level features such as edges, whereas neurons in the late stages of visual processing are responsive to more complex features such as faces. This example shows that high-level feature extraction can be readily implemented in a deep neural network.

UNSUPERVISED LEARNING

The artificial neural network in the preceding example performed abstraction as a result of supervised learning. The species of an animal in the image was explicitly given to the artificial neural network (although not to the hidden layers) during training. One might then argue that the artificial neural network performed abstraction because it was trained to do so. In other words, artificial neural networks can perform abstraction only when explicitly trained to do so in a narrow domain. For example, a network trained to classify animal species may perform poorly in other domains such as classifying vehicle types. If so, understanding learning-induced abstraction in deep artificial neural networks may provide no useful insight into the neural processes behind human abstraction. This, thankfully, is not the case. Deep neural networks can be trained to represent abstract concepts using unsupervised learning.

In 2012, the team led by Andrew Ng and Jeff Dean trained a nine-layered neural network with millions of unlabeled images (i.e., no information was provided about the images). Surprisingly, after three days of training, face-selective neurons emerged. The best-performing neuron in the network identified faces from new images with 81.7 percent accuracy.[3] This study and subsequent ones have convincingly shown that deep neural networks can represent high-level features (the prototype of face in the above example) without a specific instruction about the features of interest.

A PRIORI ABSTRACTION

More recent studies demonstrated, even more surprisingly, that deep neural networks can represent abstract concepts even without any training whatsoever, further indicating that there is something special about the deep neural network structure in representing abstract concepts. Se-Bum Paik, my colleague at the Korea Advanced Institute of Science and Technology, constructed biologically inspired deep neural networks. In one study, his team found that neurons in a simulated multilayer neural network show number-specific responses even without any training.[4] In other words, number sense emerges spontaneously from the statistical properties of multilayer neural network projections. These number-coding neurons of the simulated neural network showed similar response characteristics to those of number-coding neurons found in the brain (see chapter 9). For example, their sensitivity to discriminate between two numbers decreases as the size of the numbers increases (e.g., the sensitivity to discriminate between 1 and 2 is similar to the sensitivity to discriminate between 10 and 20), which obeys the Weber-Fechner law (the intensity of a sensation is proportional to the logarithm of the stimulus intensity).

In another study, the team found face-selective neurons in a multilayer neural network that simulated the monkey visual pathway without any training.[5] These face-selective neurons showed many response characteristics that are similar to those of face-selective neurons found in the monkey brain. These studies tell us two important points. First, number selectivity and face selectivity may emerge spontaneously in a neural network without any learning. Second, such properties are found only in neural networks with a sufficient depth (i.e., enough hidden layers). Then, a sufficiently deep neural network, such as the primate brain, may be able to generate abstract concepts (such as the number sense) without any prior experience. In other words, the brain's multilayered network structure may be the source of innate cognitive functions such as abstraction.

CONCEPT CELLS

Studies using artificial neural networks suggest that the superb human capacity for high-level abstraction may be because the human brain has more layers of connections than other animals rather than having a dedicated neural network for high-level abstraction. In fact, the basic circuit structure is similar across all mammalian cortices. From this perspective, the key to higher cognitive functions of the prefrontal cortex (see chapter 9) may be that it is the highest-order association cortex rather than the existence of a specialized neural circuitry. In particular, the frontopolar cortex is at the top of the connection hierarchy (see fig. 9.4), which may be why it deals with particularly high-level abstract concepts in humans.

But what about the hippocampus? It can also be considered the highest-order association cortex. Thus, humans may be able to imagine freely using high-level abstract concepts because those concepts are formed spontaneously as sensory inputs go through multiple layers of the neural network before reaching the hippocampus. Note that counting the number of hidden layers between sensory receptors and the hippocampus is not as straightforward as counting the number of hidden layers in a feed-forward artificial neural network because the cortex has strong recurrent projections (i.e., a group of neurons project to themselves) and one area of the cortex is typically connected with multiple areas. Also, we only have limited data for a systematic, quantitative comparison of the number of cortical hidden layers across different animal species. Nevertheless, it is clear that the human brain has a particularly large number of cortical associational areas.[6] For example, in the ventral visual stream (the pathway from the primary visual cortex to the hippocampus), the number of intervening cortical areas is way larger in humans than in mice, tree shrews, gray squirrels, and monkeys.[7]

Consider the following study, which found abstraction-related neural activity in the human hippocampus. As part of a procedure to localize the source of epileptic seizures, neural activity was recorded from the hippocampus and surrounding areas (medial temporal lobe) of human epileptic patients. Neurons in the medial temporal lobe of these patients showed diverse activity patterns in response to diverse visual stimuli. Surprisingly, some of the neurons were only responsive to specific persons or objects. For example, the neuron shown in fig. 11.3 was responsive to the pictures of Jackie Chan, an actor, but not to the pictures of other people.

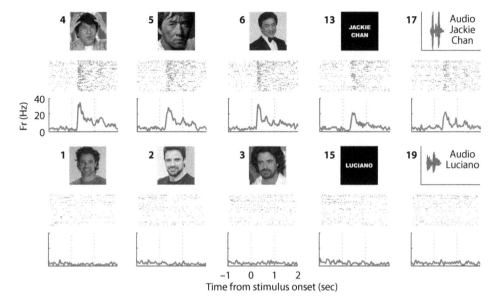

FIGURE 11.3. A human hippocampal neuron that is responsive to the actor Jackie Chan but not to another actor, Luciano Castro. Middle row, spike raster plot (each dot is a spike and each row is a trial). Bottom row, average response. Figure reproduced from Hernan G. Rey et al., "Single Neuron Coding of Identity in the Human Hippocampal Formation," *Current Biology* 30, no. 6 (March 2020): 1153 (CC BY).

Remarkably, this neuron was responsive not only to the pictures of him but also to his written and spoken names.[8] This indicates that these person-selective neurons are not simply responsive to visual features. Rather, they are responsive to a personal identity, which is an abstract concept. These cells were therefore named *concept cells*.[9]

Attempts to find similar concept cells in animals have not been successful. For example, hippocampal neurons in monkeys were independently activated by pictures and voices of other monkeys in the colony.[10] This suggests that the human hippocampus may handle particularly high-level abstract concepts, although other interpretations (e.g., abstract identity coding may not be particularly advantageous for monkeys that do not form such large social communities as humans do) are equally plausible. Notably, the prefrontal cortex and hippocampus, which are the highest-order association cortices, have direct communication channels. The hippocampus sends direct projections to the prefrontal cortex, and conversely, the prefrontal cortex sends direct projections to the hippocampus.[11] Of course, they are also connected indirectly via other brain regions such as the precuneus. It is entirely possible that the hippocampus and prefrontal cortex interact closely via these direct and indirect pathways and that such hippocampal-prefrontal cortical interactions underpin unbounded imagination using high-level abstract concepts in humans.

DEEP ARTIFICIAL NEURAL NETWORKS

If the depth of a neural network in the brain determines the level of abstract thinking, what would abstract concepts of very deep artificial neural networks be like? Nowadays, the depth of artificial neural networks routinely exceeds 100. Powerful graphics processing units, originally designed to meet the demands of modern computer games, facilitate the implementation of very deep learning algorithms. Will these deep artificial neural networks be able to form higher-level abstract concepts than human brains? Did AlphaGo, the first computer program to defeat a top-tier human Go master, prevail by using high-level abstract concepts beyond human comprehension?

It has been proposed that the human brain will reach its maximum processing capacity at approximately 3,500 cm³, which is approximately 2.5 times larger than the current one (1,350 cm³).[12] Because of energetic and neural processing constraints, a larger brain would be less efficient. It requires more energy and powerful cooling, which would increase the volume used by the blood vessels. Furthermore, neurons have inherent limits in signal processing speed, which constrains the efficiency of information processing that a brain can achieve by increasing its size. If this estimate is correct, the human brain has room to evolve into a deeper neural network. If someone is born with a deeper network structure due to a mutation, how would they think? Would their artistic sense differ from ours? If we manipulate genes so that a dog is born with neural network connections as deep as in humans, would the dog have the same capacity for abstract thinking as humans? We currently have no clear answers to these questions. However, we are not without research along this line.

DEEP REAL NEURAL NETWORKS

As elaborated in chapter 8, the current human brain is the outcome of the sudden expansion of the neocortex in the course of evolution. This is most likely due to mutations in some of the genes that control neocortex development. *ARHGAP11B* is one of the genes critical for the development of the human neocortex. It plays an essential role in the proliferation of neural stem cells during early development. In other words, it promotes the production of neurons for the neocortex. What would happen if this gene were expressed in an animal?

Thus far, scientists have expressed this gene in the mouse, ferret, and marmoset. Its expression in mice increased neocortical size as well as memory flexibility. *ARHGAP11B*-expressing mice and control mice performed similarly in remembering a fixed rewarding location. However, *ARHGAP11B*-expressing mice outperformed control mice when the rewarding location changed daily, indicating enhanced behavioral flexibility.[13] Remember from chapter 10 that people with prefrontal cortex damage have poor behavioral flexibility.

What about the marmoset, which is a small monkey? Would *ARHGAP11B* expression make the marmoset smarter? Would it allow the marmoset to think using high-level abstract concepts? As you might expect, these studies are not free from ethical issues. That's why the German and Japanese scientists who performed this work halted the project before the birth of the experimental marmoset.[14] They removed the mutant embryos surgically and examined their brains. It is therefore unknown whether these marmosets would "think" differently from ordinary ones. However, it was found that the gene expression did alter the development of the marmoset brain. Their neocortex was visibly enlarged and wrinkled (see fig. 11.4).

FIGURE 11.4. *ARHGAP11B* expression in the marmoset brain. (Left) A common marmoset. Panel reproduced from "Common Marmoset (*Callithrix jacchus*)," NatureRules1 Wiki, accessed August 28, 2022, https://naturerules1.fandom.com/wiki/Common_Marmoset (CC BY-SA). (Right top) The brain of a normal marmoset fetus. (Right bottom) The brain of a marmoset expressing human *ARHGAP11B*. Panels reproduced with permission from Michael Heide et al., "Human-Specific *ARHGAP11B* Increases Size and Folding of Primate Neocortex in the Fetal Marmoset," *Science* 369, no. 6503 (July 2020): 547.

The top of the human brain has many folds, which increase the surface area of the neocortex. This appears to be a method of harnessing an enlarged neocortex within a limited space inside the skull. By contrast, normal marmosets' brains have a smooth surface. That the altered marmoset fetuses have an enlarged neocortex with some wrinkles on the surface suggests that their "thinking," provided they were born and survived until adulthood, may differ considerably from that of normal marmosets. Of course, the brains of these mutant marmosets are still far from the complexity of the human brain. Hence, their abstract-thinking capability is unlikely to have approached that of humans. Nonetheless, with the caveat of ethical issues, this line of research may one day come up with an animal brain that is as complex as, or even more complex than, the human brain.

To summarize, the neural basis of human high-level abstract thinking is unclear. However, studies directly investigating these issues may become popular in the future, though they may be accompanied by ethical concerns.

PART IV

Beyond Imagination and Abstraction

Chapter Twelve

SHARING IDEAS AND KNOWLEDGE
THROUGH LANGUAGE

"If I have seen further it is by standing on the shoulders of giants." This famous quote in a letter from Isaac Newton to another eminent and rival scientist, Robert Hooke, is frequently used to symbolize how scientific progress is made. Discoveries, inventions, and innovations rarely occur in a vacuum. We rely on the knowledge gained by our predecessors to make intellectual progress.

Galileo Galilei, who died in the year Newton was born (1642), was one giant whose shoulders Newton stood on. Galileo laid down the foundation for understanding the motion of an object, including the formulation of the concept of *inertia* (an object remains in the same state of motion unless an external force is applied), which greatly influenced Newton. Johannes Kepler was another giant for Newton. Kepler, using the astronomical data Tycho Brahe collected over decades, discovered laws of planetary motion. Building on Kepler's work, Newton developed general laws of motion that can explain all motion, be it the orbit of the Earth around the Sun or an apple falling from a tree. Before Newton was born, Galileo and Kepler were able to see further than their contemporaries because they stood on the shoulders of Nicolaus Copernicus, who developed a heliocentric (Sun-centered) model of the solar system.

We have, thus far, considered neural mechanisms behind an individual's thought processes enabling innovation, focusing on imagination and abstraction. But we have not paid much attention to another critical faculty of the human mind for innovation, namely the capacity to share ideas and knowledge. For the advancement of science and technology, such as those shown in fig. 0.1, social communication involving high-level abstract concepts is essential. It is difficult to imagine a human civilization as complex and advanced as today's without language. In this respect, acquiring linguistic capacity represents an important step in human evolution, as described by Mark Pagel: "Possessing language, humans have had a high-fidelity code for transmitting detailed information down the generations. The information behind these was historically coded in verbal instructions, and with the advent of writing it could be stored and become increasingly complex. Possessing language, then, is behind humans' ability to produce sophisticated cultural adaptations that have accumulated one on top of the other throughout our history as a species."[1]

In this chapter, we will explore language, a unique form of human communication, which elevated mankind's collective creativity to a new dimension. We will examine a few selected topics of language, including its neural basis, its relationship with abstract thinking, and its origin.

UNIVERSALITY OF LANGUAGE

Language is a system by which sounds, symbols, and gestures are used according to grammatical rules for communication. No mute tribe has ever been discovered, indicating that language is universal in all human societies. Although there are many different languages (more than seven thousand languages and dialectics throughout the world), all languages are used in similar ways—to elaborate personal experiences, to convey subtle emotions, to solicit information (ask questions), and to make up stories (sometimes for deception), along with many other facets of human communication. The process of language acquisition is also similar across all cultures. Moreover, there is a critical period during which children learn to use a language effortlessly without any formal training if they are exposed to a normal linguistic environment. Pagel further explains, "All human languages rely on combining sounds or 'phones' to make words, many of those sounds are common across languages, different languages seem to structure the world semantically in similar ways, all human languages recognize the past, present and future and all human languages structure words into sentences. All humans are also capable of learning and speaking each other's languages."[2] The universality of language suggests that the human brain has evolved special language-processing systems.

THE UNIQUENESS OF HUMAN LANGUAGE

Language is often placed at the top of the list for the criteria that separate humans from other animals. Although other animals besides humans communicate with each other, it is doubtful that animal communication can be considered language. The following commonly mentioned features of human language are either absent or substantially limited in the other animals' communications:[3]

Discreteness: Human language consists of a set of discrete units. The smallest unit of speech is a phoneme (or a word sound). Forty-four discrete phonemes in English distinguish one word from another. For example, the words *cat* and *can* share the first two phonemes, *c* and *a*, but differ by the last one. At another level, individual words are combined to construct phrases and sentences. There are sentences made up of five or six words, but there are no sentences made up of five and a half words. *Cat* and *can* are distinctly different words, and there is no meaning in-between them.

Grammar: Rules govern the underlying structure of a language. Linguistic units are combined according to these rules to express meanings.

Productivity: There is no upper bound to the length of a sentence one can construct with words. Humans have the ability to produce an infinite number of expressions by combining a limited set of linguistic units.

Displacement: Humans can talk about things that are not immediately present (e.g., past and future events) or that do not even exist (e.g., imaginary creatures on Mars).

We combine discrete units of language (e.g., words) according to grammatical rules to generate indefinite utterances (e.g., sentences) that include things that do not even exist. Many animal species are capable of sophisticated communication that shows some of the features of human language. For example, the waggle dance of a honeybee describes the location and richness of a food source,[4] and prairie dog alarm calls vary according to predator type and individual predator characteristics including color and shape.[5] However, no animal communication system displays all the features of human language. Human language is a complex, flexible, and powerful system of communication. The creative use of words according to rules can generate unlimited expressions. No other animal communication comes close to human language.

CAN ANIMALS BE TAUGHT HUMAN LANGUAGE?

Animals' natural communication falls short of human language, but is it possible to teach animals to use language? The question is whether other animals can be trained to combine symbols in original ways according to some grammatical system to express new things. There have been numerous attempts to teach animals (such as chimpanzees and dolphins) human language, and the results are controversial. Let's examine the case of Koko (July 4, 1971–June 19, 2018), a female western lowland gorilla that learned and lived in a language environment (sign language and spoken English) from the age of one. It has been claimed that Koko was able to combine vocabularies in "creative and original sign utterances."[6]

> Koko has generated numerous novel signs without instruction, modulated standard signs in ASL (American Sign Language) to convey grammatical and semantic changes, used signs simultaneously, created compound names (some of which may be intentional metaphors), engaged in self-directed and noninstrumental signing, and has used language to refer to things removed in time and space, to deceive, insult, argue, threaten, and express her feelings, thoughts, and desires. These findings . . . support the conclusion that language acquisition and use by gorillas develops in a manner similar to that of human children.[7]

The following is a conversation in sign language between the trainer, Francine Patterson (denoted as "F"), and Koko (denoted as "K") about Koko's favorite kitten, All Ball, who died in a car accident a few months earlier:[8]

F: How did you feel when you lost Ball?
K: WANT.
F: How did you feel when you lost him?
K: OPEN TROUBLE VISIT SORRY.
F: When he died, remember when Ball died, how did you feel?
K: RED RED RED BAD SORRY KOKO-LOVE GOOD.

This was interpreted as Koko's expression of sad feelings about the death of All Ball. However, skeptical scientists suspect whether Koko, and other language-taught animals as well, were truly able to combine symbols in creative ways. We cannot exclude the possibility that these animals learned to act in certain ways under certain circumstances (simple stimulus-response associations) to get a reward without understanding the meaning behind their actions. For example, they may have learned to make certain actions in response to the trainer's subtle unconsciousness cues. This is commonly called the "Clever Hans effect":

> In the first decade of the 20th century, a horse named Hans drew worldwide attention in Berlin as the first and most famous "speaking" and thinking animal. . . . . Hans performed arithmetic operations precisely, tapping numbers with his hoof, and answered questions in the same way, using an alphabet with letters replaced by numbers (A = 1, B = 2, C = 3 . . .) on a blackboard in front of his eyes. Thus Hans combined letters to words, words to sentences, and sentences to thoughts and ideas. . . . Professor Oscar Pfungst . . . found that the horse was unable to answer any question if the questioning person did not know the answer. . . . Furthermore, the horse was unable to answer any question when a screen was placed such that it could not see the face of his examiner. . . . It turned out that the horse was an excellent and intelligent observer who could read the almost microscopic signals in the face of his master, thus indicating that it had tapped or was about to tap the correct number or letter and would receive a reward. In the absence of such a signal, he was unable to perform. Indeed, Pfungst himself found that he was unable to control these clues as the horse continued to answer correctly when his face was visible to it.[9]

Another concern is that Koko's signs had to be interpreted by the trainer. "Conversations with gorillas resemble those with young children and in many cases need interpretation based on context and past use of the signs in question. Alternative interpretations of gorilla utterances are often possible."[10] A sequence of hand signs that were interpreted as "you key there me cookie" by the trainer (supposedly meaning "Please use your key to open that cabinet and get out a cookie for me to eat") may actually represent Koko's "flailing around producing signs at random in a purely situation-bound bid to obtain food from her trainer, who was in control of a locked treat cabinet. . . . Koko never said anything: never made a definite truth claim, or expressed a specific opinion, or asked a clearly identifiable question. Producing occasional context-related signs, almost always in response to Patterson's cues, after years of intensive reward-based training, is not language use."[11]

To summarize, it is still unclear whether animals other than humans can learn to use a language. Even though we may admit the possibility, their use of words is different from that of adult humans. At best, the animal's acquired linguistic capacity is not better than a three-year-old child.

LANGUAGE AREAS OF THE BRAIN

It seems clear that the complex and flexible system of language humans use is unique. Because of the lack of a legitimate animal model, studying details of the neural processes essential to language is limited. We have made enormous progress in understanding the neural processes underlying many brain functions that are shared by humans and other animals. For example, we can explain learning and memory in terms of altered neural circuit dynamics as a consequence of strengthened neural connections (enhanced synaptic efficacy; see chapter 4). Furthermore, we have identified molecular processes underlying synaptic efficacy change. Scientists can even manipulate specific neurons (artificially turn on and off specific neurons using a technique called optogenetics) to implant a false memory into a mouse.[12] Compared to our understanding of the neural basis of memory, our understanding of the neural basis of language is far behind. Nevertheless, we have identified regions of the human brain that are specialized in language processing.

The first identified area specialized in language processing is Broca's area. In 1861, a French neurologist, Paul Broca, described a patient who lost the ability to speak without other physical or intellectual deficits. The patient's name was Louis Victor Leborgne, but he was known as Tan because "tan" was the only syllable he could utter. A postmortem examination revealed a large lesion in the lower middle of the left frontal lobe (the left inferior frontal gyrus), the area now known as Broca's area (see fig. 12.1).[13] People with damage in that area have relatively intact comprehension but have great difficulty speaking. Their speech is often telegraphic, consisting mostly of nouns and verbs. This type of aphasia is called Broca's, nonfluent, or motor aphasia.

A few years later, a German neurologist, Carl Wernicke, observed a different type of aphasia. These patients can speak fluently but have difficulty in comprehension. Even though their speech is fluent (rhythmic and grammatical), it does not make sense, and they are often unaware of it. It appears their speech production system operates without proper control over the contents. In these patients, damage is found in the posterior part of the left

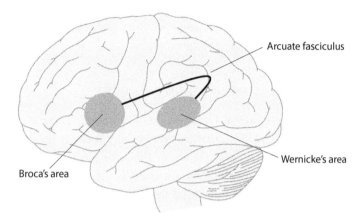

FIGURE 12.1. Broca's and Wernicke's areas. In the Wernicke-Geschwind model, these areas oversee language production and comprehension, respectively, and they exchange information via the arcuate fasciculus, which is the major axon bundle connecting them. When we repeat spoken words, speech sounds, first processed by the auditory system, are converted into meaningful words in Wernicke's area. The word signals are then sent to Broca's area (via the arcuate fasciculus) where they are converted into motor signals to move muscles for speech production. When we read written text, visual signals, initially processed by the visual system, are transformed into pseudo-auditory signals and sent to Wernicke's area. Wernicke's area then converts those signals into meaningful words.

temporal lobe near the primary auditory cortex, the area now known as Wernicke's area (see fig. 12.1).[14] This type of aphasia is called Wernicke's, fluent, or sensory aphasia.

Wernicke proposed an early neurological model of language that was later extended by Norman Geschwind. Although the Wernicke-Geschwind model has been enormously influential, it is based on oversimplifications and is incorrect in many respects.[15] For instance, there are no sharp functional distinctions between Broca's and Wernicke's aphasia; most aphasias involve both comprehension and speech deficits. Some argue that "the model is incorrect along too many dimensions for new research to be considered mere updates."[16] Moreover, language processing involves multiple levels of processing (phonological, syntactic, and semantic) and interactions with multiple neural systems (such as motor, memory, and executive control systems). Hence, framing language processing merely as comprehension and production would not be useful in revealing diverse neural processes underlying language. More elaborate neurological models of language now focus on functional subcomponents of language processing rather than a comprehension versus production distinction. We are not going to examine these new models here, but the field is moving "from coarse characterizations of language and largely correlational insights to fine-grained cognitive analyses aiming to identify explanatory mechanisms."[17]

LANGUAGE AND THOUGHT

The Irish poet and playwright Oscar Wilde once wrote, "There is no mode of action, no form of emotion, that we do not share with the lower animals. It is only by language that we rise above them, or above each other—by *language, which is the parent, and not the child, of thought.*"[18]

I began this chapter by emphasizing the role of language in sharing and accumulating knowledge across generations—the communication aspect of language. From this standpoint, language should be considered the child of thought—the means of expressing our inner thoughts. However, according to Wilde, language not only expresses but also shapes our thoughts, and this view has been shared by many renowned figures such as Ludwig Wittgenstein ("The limits of my language mean the limits of my world")[19] and Bertrand Russell ("Language serves not only to express thoughts, but to make possible thoughts which could not exist without it").[20]

No one would disagree that language influences thought. There is no doubt that "language use has powerful and specific effects on thought. That's what it is for, or at least that is one of the things it is for—to transfer ideas from one mind to another mind."[21] Most of us hear an "inner voice" or "inner speech" while thinking. However, is language necessary for thought? If we lose language, do we lose thought as well? The answer to this question is clear: "Not really." Neuroscientific studies indicate that the brain areas linked to language are distinct from those linked to other cognitive functions such as memory, reasoning, planning, decision-making, and social cognition. Also, people with global aphasia, meaning they lost their linguistic capacity due to brain damage, can perform a variety of tasks requiring abstract reasoning.[22] Furthermore, as we discussed in chapter 8, we are not the only animal species capable of abstract reasoning; chimpanzees and dolphins, for example, show impressive abstract inference capabilities. It is thus clear that language is not a prerequisite for thought. Many people come up with creative ideas through nonverbalized thinking. Albert Einstein claimed that imagining he was chasing a beam of light played a memorable role in developing the theory of special relativity. In fact, some people think without internal speech.

Although thinking without words is possible, it does not necessarily indicate that language has no role in shaping our thoughts. There has been much debate on whether and how language influences thought. Some argue that the primary role of language is limited to communication. Language enables the expression of inner thoughts but is not necessary for abstract thought itself. According to the influential *language of thought hypothesis*, we think using a mental language (often called "mentalese") that has a linguistic structure. In other words, thoughts are really "sentences in the head."[23] From this standpoint, it is unlikely that the influence of language on thought goes beyond that through its role in communication:

> Language is a lot like vision. Language and vision are both excellent tools for the transfer of information. People who are blind find it harder to pick up certain aspects of human culture than people who can see, because they lack the same access to books, diagrams, maps, television, and so on. But this does not mean that vision makes you smart, or that explaining how vision evolves or develops is tantamount to explaining the evolution and development of abstract thought. Language may be useful in the same sense that vision is useful. It is a tool for the expression and storage of ideas. It is not a mechanism that gives rise to the capacity to generate and appreciate these ideas in the first place.[24]

Contrary to this view, many people consider language as something more than a vehicle for the communication of thought. A strong version of this perspective, known as the Sapir–Whorf hypothesis or the *linguistic relativity hypothesis*, argues that our perception of the world is determined by the structure of a specific language we use.[25] According to this hypothesis, Chinese and English speakers may have different ways of thinking. This hypothesis, at least in its radical version, is not widely accepted. However, many researchers believe that language can exert profound influences on our thoughts, especially those involving abstract concepts, beyond its role in communication.

For example, Guy Dove views language as a form of *neuroenhancement*, stating that "the neurologically realized language system is an important subcomponent of a flexible, multimodal, and multilevel conceptual system. It is not merely a source for information about the world but also a computational add-on that extends our conceptual reach."[26] Anna Borghi emphasizes the role of language in representing abstract concepts: "The potentialities of language are maximally exploited in representing abstract concepts:

labels can help us to glue together the heterogeneous members of abstract concepts, and inner speech can improve the capability of our brain to track information on internal states and processes and to introspectively look at ourselves."[27]

According to this perspective, language is not merely used for exchanging ideas but also plays a role in representing abstract concepts such as *liberty* and *integrity* (but not much in representing concrete concepts such as *table* and *cat*). In this regard, Lupyan and Winter noted that language is way more abstract than iconic, even though iconicity—the resemblance of a word form to its meaning (*tweet*, *click*, and *bang* are examples of iconic words)—can be immensely useful in language learning and communication.[28] According to them, "abstractness pervades every corner of language" and "the best source of knowledge about abstract meanings may be language itself."[29] We learn many abstract concepts by learning language rather than through actual interactions with the world.[30]

People with aphasia show no impairment (or only weak impairment) in abstract reasoning.[31] This argues against the claim that language should be considered a "neuroenhancement" that extends our conceptual reach.[32] However, language deprivation in childhood (deaf children born to hearing parents are often insufficiently exposed to sign language in their first few years of life) can lead to "language deprivation syndrome," which involves deficits in a variety of cognitive functions, such as understanding cause and effect and inferring others' intentions.[33] Brain imaging studies revealed changes in cortical architecture and activation patterns in these children.[34] Intriguingly, some observations point to a global deficit in their abstraction abilities. "People with language deprivation do not excel at such seemingly nonverbal (but clearly conceptual) skills as playing chess or doing math . . . They can struggle to name basic feelings, to recognize the social boundaries that are encoded in words such as 'girlfriend,' or to reflect on their own experiences. They struggle to see patterns and make generalizations. Lessons learned in one arena are not easily applied to others."[35]

Thus, while available evidence is limited, language may play a role in expanding and refining abstract thinking during the critical period. Language may play a scaffolding role—"essential in the construction of a cognitive architecture, but once the system achieves a level of stability, the linguistic scaffolding can be removed."[36] Language may be one of the critical factors during development that pushes the boundary of thought deep into the abstract domain, allowing us to think seamlessly using both concrete and abstract concepts, which sometimes leads to a category mistake (see chapter 8). Again, the evidence is limited. What we speculated about the potential role of language in abstraction may or may not be correct. The jury is still out.

ORIGIN OF LANGUAGE

Even though uncertainty remains regarding the language-thought relationship, there is no doubt that the acquisition of linguistic capacity was a monumental step in human evolution. How have we acquired this capacity? What is the origin of language? This is a critical issue to understand how we have become what we are—*Homo innovaticus*. The origin of language has been debated for a long time among linguists, anthropologists, philosophers, psychologists, and neuroscientists. There are many perspectives, some of which are in conflict with one another. Unfortunately, that indicates that we are still far from fully understanding this matter.

Understanding the origin of language is a tough challenge for several reasons. First, language is unique to humans. No other animal communication system comes close to the complex, flexible, and powerful system humans use for communication. Consequently, unlike many other biological functions, it is difficult to gain insights from comparative approaches (i.e., examining and comparing language functions in diverse animal species). It is also difficult to perform invasive studies (such as recording neural activity with a microelectrode or inactivating a localized brain area) due to the lack of an animal model.

Second, we lack direct fossilized records of language. It is difficult to relate archaeological records directly to the evolution of language. Likewise, brain endocasts can tell us little about language function. Fossil records of speech organs, such as the vocal tract, could be related to the evolution of speech, but there is a huge gap between the capacity for producing articulated sounds and the capacity for grammatically combining symbols to generate novel expressions. The laryngeal descent theory has been popular for the past fifty years. This theory contends that laryngeal descent occurred in our ancestors, but not in other primates, around two hundred thousand years ago, allowing for the production of contrasting vowel sounds, the development of speech, and finally the development of language in *Homo sapiens*. However, a recent comprehensive review concludes that "evidence now overwhelmingly refutes the longstanding laryngeal descent theory, which pushes back 'the dawn of speech' beyond ~200 ka ago to over ~20 Ma ago."[37]

Third, the dynamics of social interaction must be considered in explaining the evolution of language. It is often less straightforward to explain the evolution of a social trait, such as altruism, compared to a nonsocial trait. Perhaps humans are the only animals showing *true* altruism. We often help strangers without expecting reciprocal benevolence. We even take personal loss to punish norm violators (altruistic punishment). These human behaviors are not readily explained in terms of biological fitness (the ability to pass on our own genes to the next generation).[38] Obviously, altruism enhances the survival fitness of a group. However, within a group, the survival fitness would be higher among selfish individuals. Hence, it is not clear how altruistic traits were selected over selfish ones across generations. Likewise, being able to share information would be a big advantage for a group. However, it is unclear how this trait initially appeared and survived across generations. Language is useless if you are the only one who can speak.

Here, instead of a comprehensive overview of theories on the origin of language, we will selectively examine a few only to the extent to reveal the diversity of voices in the field. One of the debates related to the origin of language is whether language evolved suddenly or gradually. Recall the debate over the evolution of behavioral modernity and the human revolution (see chapter 10). Noam Chomsky proposed that modern language evolved suddenly as a consequence of a fortuitous chance mutation.

> Human capacities for creative imagination, language and symbolism generally, mathematics, interpretation and recording of natural phenomena, intricate social practices, and the like, a complex of capacities that seem to have crystallized fairly recently, perhaps a little over 50,000 years ago, among a small breeding group of which we are all descendants—a complex that sets humans apart rather sharply from other animals, including other hominids, judging by traces they have left in the archaeological record. . . . The invention of language . . . was the "sudden and emergent" event that was the "releasing stimulus" for the appearance of the human capacity in the evolutionary record—the "great leap forward' as Jared Diamond called it, the results of some genetic event that rewired the brain, allowing for the origin of modern language with the rich syntax that provides a multitude of modes of expression of thought, a prerequisite for social development and sharp changes of behavior that are revealed in the archaeological record[39]

The discovery of the gene *FOXP2* raised the hope that we might be able to identify the "fortuitous mutation" Chomsky proposed. *FOXP2* was identified by the 2001 genetic analysis of a British family known as KE; half of its members were severely impaired in speech production.[40] Many of the affected KE family members have low IQs, but some of them show language problems with normal IQs, indicating that the language problem is not because of a general cognitive impairment. A subsequent comparative genetic study has shown that the human *FOXP2* protein differs from those of other great apes only by two amino acids.[41] This study also found evidence for "selective sweep," a process in which a rare beneficial mutation rapidly increases in frequency by natural selection. These findings suggest that a beneficial mutation in the *FOXP2* gene, estimated to have happened about two hundred thousand years ago, may have triggered the development of modern language in humans. However, contrary to the initial report, a later study found no evidence for a selective sweep in *FOXP2*.[42] Moreover, an ancient DNA sequencing study revealed that Neanderthals share the evolutionary changes in *FOXP2* (two amino acids that differ from other great apes) with modern humans.[43] Thus, the beneficial changes in *FOXP2* must have happened before the split of our ancestors from Neanderthals (about half a million years ago), but there is no clear evidence for language use by Neanderthals. Some advocate language use by Neanderthals, but many are skeptical about this possibility.[44]

Currently, most researchers seem to favor the idea that human language evolved in steps over a long period, just like most biological functions did, even though it is unclear how exactly this gradual process unfolded during human evolution. The following is one hypothetical scenario for the steps of language evolution:

> In an early stage, sounds would have been used to name a wide range of objects and actions in the environment, and individuals would be able to invent new vocabulary items to talk about new things. In order to achieve a large vocabulary, an important advance would have been the ability to "digitize" signals into sequences of discrete speech sounds—consonants and vowels—rather than unstructured calls. This would require changes in the way the brain controls the vocal tract and possibly in the way the brain interprets auditory signals. . . . A next plausible step would be the ability to string together several such "words" to create a message built out of the meanings of its parts. This is still not as complex as modern language. It could have a

rudimentary "me Tarzan, you Jane" character and still be a lot better than single-word utterances. In fact, we do find such "protolanguage" in two-year-old children, in the beginning efforts of adults learning a foreign language, and in so-called "pidgins," the systems cobbled together by adult speakers of disparate languages when they need to communicate with each other for trade or other sorts of cooperation. . . . A final change or series of changes would add to "protolanguage" a richer structure, encompassing such grammatical devices as plural markers, tense markers, relative clauses, and complement clauses ("Joe thinks *that the earth is flat*").[45]

Another passionately debated issue regarding the origin of language is the relationship of language to other cognitive functions. Some argue that the human language system is a stand-alone system separated from all other mental faculties, while others argue that language draws on other cognitive functions. Again, let's examine a few of these theories.

Michael Tomasello is one of the researchers advocating the idea that language is a manifestation of more general cognitive functions. He argues that humans and great apes both have the ability to comprehend the intentions of others. However, only humans engage in activities involving *joint* intentions and attention, which he named "shared intentionality": "In addition to understanding others as intentional, rational agents, humans also possess some kind of more specifically social capacity that gives them the motivation and cognitive skills to feel, experience, and act together with others—what we may call, focusing on its ontogenetic endpoint, shared (or 'we') intentionality. As the key social-cognitive skill for cultural creation and cognition, shared intentionality is of special importance in explaining the uniquely powerful cognitive skills of *Homo sapiens*."[46]

According to Tomasello, only humans have the cognitive capacity to represent a shared goal (e.g., you and I work together to open a box) and joint intention (e.g., you hold the box while I cut it open). The capacity to represent shared intentionality begins to emerge around the first birthday when children start using words. He argues that unique human cognitive capacities, such as language use and theory of mind, are *derived* from the understanding and sharing of intentions (joint intentionality).

Saying that only humans have language is like saying that only humans build skyscrapers, when the fact is that only humans (among primates) build freestanding shelters at all. Language is not basic; it is derived. It rests on the same underlying cognitive and social skills that lead infants to point to things and show things to other people declaratively and informatively, in a way that other primates do not do, and that lead them to engage in collaborative and joint attentional activities with others of a kind that are also unique among primates.[47]

Steven Pinker is among the researchers advocating the opposite view, namely that language is fundamentally different from other types of cognition. He argues that language evolved for the *communication* of complex propositions.

> What is the machinery of language trying to accomplish? The system appears to have been put together to encode propositional information—who did what to whom, what is true of what, when, where and why—into a signal that can be conveyed from one person to another. It is not hard to see why it might have been adaptive for a species with the rest of our characteristics to evolve such an ability. The structures of grammar are well suited to conveying information about technology, such as which two things can be put together to produce a third thing; about the local environment, such as where things are; about the social environment, such as who did what to whom, when, where and why; and about one's own intentions, such as *If you do this, I will do that,* which accurately conveys the promises and threats that undergird relations of exchange and dominance.[48]

Pinker contends that humans, unlike other animals, evolved to perform cause-and-effect reasoning and rely on information for survival. It would then be advantageous for them to come up with a means to *exchange* information. He argues that "three key features of the distinctively human lifestyle—knowhow, sociality, and language—co-evolved, each constituting a selection pressure for the others."[49]

According to Marc Hauser, Noam Chomsky, and Tecumseh Fitch, a distinction should be made between the faculty of language in the broad sense (FLB) and in the narrow sense (FLN). The FLN includes only the core computational mechanisms of recursion (a clause can contain another clause so that one can generate an infinite number of sentences of any size, such as "I wonder if he knows that she believes that we suspect him as the criminal") and the FLB has two additional components—a sensory-motor system and a conceptual-intentional system. Together with the FLN, they provide "the capacity to generate an infinite range of expressions from a finite set of elements" (discrete infinity).[50]

They hypothesize that the FLN is recently evolved, unique to humans, and unique to language, so "there should be no homologs or analogs in other animals and no comparable processes in other domains of human thought."[51] By contrast, many aspects of the FLB are shared with other vertebrates. They also speculate that the FLN (recursion mechanism) may have evolved "to solve other computational problems such as navigation, number quantification, or social relationships."

> One possibility, consistent with current thinking in the cognitive sciences, is that recursion in animals represents a modular system designed for a particular function (e.g., navigation) and impenetrable with respect to other systems. During evolution, the modular and highly domain-specific system of recursion may have become penetrable and domain-general. This opened the way for humans, perhaps uniquely, to apply the power of recursion to other problems. This change from domain-specific to domain-general may have been guided by particular selective pressures, unique to our evolutionary past, or as a consequence (by-product) of other kinds of neural reorganization.[52]

I mentioned the social nature of language as a challenge to understanding its origin. One of the theories we examined, Tomasello's shared intentionality theory (sharing goals and intentions with others), deals with this issue, but it's not the only one. Numerous other theories focus on the social nature of language. To examine one more, the ritual/speech coevolution theory argues that we must consider the whole human symbolic culture to understand the origin of language. Advocates of this theory propose that language works only after establishing high levels of trust in society because words are cheap (low-cost) signals and therefore inherently unreliable. Chris Knight suggests that "any increase in the proportion of trusting listeners increases the rewards to a liar, increasing the frequency of lying. Yet until hearers can safely assume honesty, their stance will be indifference to volitional signals. Then, even lying will be a waste of time. In other words, there is a threshold of honest use of conventional signals, below which any strategy based on such signaling remains unstable. To achieve stability, the honest strategy has to predominate decisively over deception."[53] Therefore, in order for verbal communication to be possible, "what is at stake is not only the truthfulness of reliability of particular messages but credibility, credence and trust themselves, and thus the grounds of the trustworthiness requisite to systems of communication and community generally."[54]

Knight further describes how this theory highlights the role of religion and ritual in establishing social trust: "Words are cheap and unreliable. Costly, repetitive and invariant religious ritual is the antidote. At the apex is an 'ultimate sacred postulate'—an article of faith beyond possible denial. Words may lie, so it is claimed, but 'the Word' emanates from a higher source. Without such public confidence upheld by ritual action, faith in the entire system of interconnected symbols would collapse. During the evolution of humanity, the crucial step was therefore the establishment of rituals capable of upholding the levels of trust necessary for linguistic communication to work."[55]

Thus, to understand the origin of language, we must consider a wider context—the human symbolic culture as a whole—of which language is one integral component, according to this theory.

MIRROR NEURON HYPOTHESIS

As can be seen from these examples, there exist widely divergent views on the origin of language that span multiple disciplines. Before wrapping up this chapter, let's examine one more theory that was derived from neuroscientific research: the mirror neuron hypothesis.

What are mirror neurons? They are neurons found in the monkey premotor cortex that are activated by both performing and observing an action. The premotor cortex is located between the prefrontal cortex and primary motor cortex. The prefrontal cortex plans a goal-directed behavior (discussed in chapter 9) and the primary motor cortex controls individual muscle contractions. The premotor cortex is thought to play an in-between role of selecting an appropriate motor plan to achieve a desired outcome (e.g., grasping an object). Roughly speaking, the prefrontal, premotor, and primary motor cortices are in charge of strategies, tactics, and execution, respectively, of goal-directed behavior.

In the 1990s, Giacomo Rizzolatti and colleagues discovered that some neurons in the monkey premotor cortex fire when the monkey is performing a certain action (such as grasping an object) as well as observing the same action performed by another (see fig. 12.2).[56] A later study showed that not only visual (seeing) but also auditory (hearing) signals are effective in

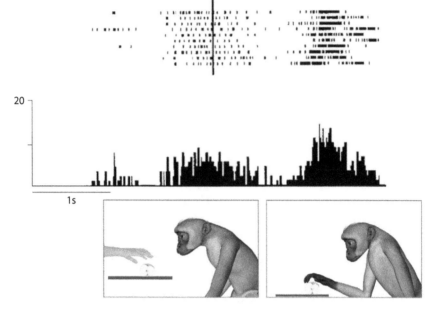

FIGURE 12.2. A mirror neuron recorded from the monkey premotor cortex. This neuron elevated its activity in response to the observation of the experimenter's grasping action as well as the monkey's own grasping action. Figure reproduced from Antonino Errante and Leonardo Fogassi, "Functional lateralization of the mirror neuron system in monkey and humans," *Symmetry* 13, no. 1 (2021): 77 (CC BY).

activating some mirror neurons (audiovisual mirror neurons) in the monkey premotor cortex.[57] For example, a neuron in the premotor cortex fired when the monkey broke a peanut (doing), when the monkey observed a human breaking a peanut (seeing), and when the monkey heard the peanut being broken without seeing the action (hearing). This area of the premotor cortex in the monkey brain (area F5) is thought to be homologous to Broca's area in the human brain, and Broca's area also increases activity during both performing and observing a finger movement.[58]

These observations led to the proposal that the mirror neuron system (or "observation/action matching" system) may have served as the basis for the evolution of language.[59]

Mirror neurons in monkey premotor cortex provide the primate motor system with an intrinsic capacity to compare the actions it generates with those generated by other individuals. It has therefore the potential to be both the sender and the receiver of message. . . . The problem of transforming an underlying case structure to a sentence (one of many possibilities) which expresses it may be seen as analogous to the nonlinguistic problem of planning the right order of actions to achieve a complex goal. Suppose the monkey wants to grab food, but then—secondarily—realizes it must open a door to get the food. Then in planning the action, the order food-then-door must be reversed. This is like the syntactic problem of producing a well-formed sentence, but now viewed as going from "ideas in the head" to a sequence of words/actions in the right order to achieve some communicative or instrumental goal.[60]

This theory is in line with the view that the origin of language is gestural communication rather than animal calls.[61] Note that the mirror neuron system alone is insufficient to support the syntax of human language, especially hierarchical recursion. According to Michael Arbib and Mihail Bota, linking Broca's area to the prefrontal cortex planning system allows us to "assemble verb–argument and more complex hierarchical structures, finding the words and binding them correctly."[62]

In summary, there is no doubt that language is a critical element of human innovation. The acquisition of linguistic capacity was a truly monumental step in human evolution. Language allows the exchange of ideas and the accumulation of knowledge across generations. Language may also play a role in setting up proper abstract-thinking capacity during development. Undoubtedly, understanding the origin of language is critical for understanding how we have become *Homo innovaticus*. Unfortunately, the origin of language is far from being clear, partly due to the uniqueness of human language. No other animal communication comes close to human language in terms of flexibility, complexity, and power. Its uniqueness, together with the paucity of archaeological evidence and its social nature, poses challenges to studying the origin of language. We have many theories but are yet to draw a robust consensus on how the faculty of language has evolved; "The richness of ideas is accompanied by a poverty of evidence."[63] Nevertheless, the origin of language draws interest from multiple disciplines. Many scholars have attempted to solve this problem by standing on the shoulders of previous great thinkers, which is made possible because of language.

Chapter Thirteen

ON CREATIVITY

Imagination is a fundamental aspect of human nature. We imagine the realistic as well as the unrealistic when we can afford not to pay attention to the outside world. This is believed to be an adaptive process. I argued in chapters 6 and 7 that imagination may be rooted in the process of simulation-selection or *offline reinforcement learning*, whereby we reinforce high-value options in diverse circumstances we may encounter in the future. By mentally exploring different options and evaluating their potential value, we are better prepared to make optimal decisions in real-life situations. This may be one reason why we are drawn to fiction and dramas; they offer a platform to exercise our imagination and engage in this simulation-selection process. Unlike other skills, imagination is an innate ability that doesn't need practice or training. It is as natural and effortless as breathing.

How about creativity? Is it possible to boost one's creativity? If so, how? Since this book is not intended to expand on the neuroscience of creativity in a comprehensive manner, I suggest interested readers look for these books by Anna Abraham and Rex Jung and Oshin Vartanian.[1] The first is a general and balanced introduction to the neural basis of creativity, and the second is a collection of writings by leading scientists in the field. Even though creativity is not the primary focus of this book, imagination and creativity are inextricably linked. Imagination does not necessarily yield creative ideas. Nor is imagination always necessary to generate creative ideas. However, creative ideas are often generated out of imagination. We examined potential neural processes underlying imagination during idle states (conscious rest and sleep) in chapters 3 and 6. In this chapter, we will try to relate what we have examined in the previous chapters to creativity. We will also examine the practical issue of how to become creative.

SPONTANEOUS CREATIVITY

Creativity refers to the capacity to yield new and valuable outcomes, such as scientific theories, artistic works, technologies, and solutions to various problems. There are two important elements of creativity: originality (novelty) and effectiveness (value, appropriateness, fit, or meaningfulness). A third factor is surprise, which is related to how unlikely it is that a given idea will be generated. An idea that is both original and useful but not surprising at all may not be considered creative.[2] Creativity is highly relevant to imagination because imagination often yields original ideas. A study has shown that hippocampal damage, which impairs vivid imagination (see chapter 1), also disrupts creative thinking.[3] Exuberant imagination, on the other hand, does not necessarily lead to a high level of creativity. This is because not all imaginations are of high value; in fact, imagination itself is value neutral. We may use our imagination to think about something totally useless or even antisocial.

Creativity is not a unitary faculty of the mind. It is "a complex and multifaceted construct" that can manifest in many ways.[4] Currently, there is no general consensus on how to divide creativity into subtypes. However, one may distinguish creativity according to its domain. For example, artistic creativity is different than scientific creativity. Albert Einstein and Pablo Picasso are considered highly creative people in science and art, respectively, but not in other fields.

Here, we will adopt the classification proposed by Arne Dietrich.[5] He distinguished deliberate versus spontaneous modes of information processing for creativity. We may come up with a novel solution to a problem at hand by focused reasoning. This type of creativity is referred to as *deliberate creativity*. It is believed that the prefrontal cortex (see chapter 9) plays a critical role in this process. In contrast, we often come up with creative ideas when we are relaxed without focusing on a particular problem or even during sleep. This type of creativity is referred to as *spontaneous creativity*. The default mode network is thought to be critical for spontaneous creativity.

Note that deliberate and spontaneous processes do not work independently for creativity. Rather, creative thinking usually involves a dynamic interplay between controlled and spontaneous processes. In terms of brain dynamics, creative thinking is thought to involve dynamic interactions among multiple large-scale neural networks (see chapter 10). In particular, dynamic interactions between the default mode network and the central executive network (or frontoparietal network, which includes the prefrontal cortex) are observed during diverse creative thinking tasks.[6] The default mode network is thought to be associated with the spontaneous generation of ideas, while the central executive network is thought to oversee the evaluative process that constrains ideas to meet specific task goals. Here, to relate creativity to the primary thesis of this book—unconstrained imagination during idle states—we will focus only on spontaneous creativity.

MEMORY AND CREATIVITY

Let's recap the neural underpinning of imagination. The default mode network is activated in association with internal mentation during idle states (see chapter 1). This occurs when the hippocampus is synchronized with the neocortex and generates diverse activity sequences. Studies indicate that changes in neural connections take place in the hippocampus during the storage of new experiences as memories (see chapter 4). Later, when we can afford not to pay attention to the outside world, hippocampal inhibitory control is weakened, and the hippocampal neural network (especially the CA3 neural network) generates diverse activity sequences. During this process, the hippocampal neural network generates not only experienced but also unexperienced activity sequences (see chapter 3). This way, the hippocampal neural network supports both memory retrieval and imagination.

An important point here is that the content of imagination is not free from the content of memory. Because memory encoding accompanies changes in neural connections to facilitate the reactivation of experienced sequences, the hippocampal neural network inevitably generates activity sequences that are related to remembered (i.e., experienced) sequences. In other words, the hippocampus tends to regenerate activity sequences that are identical (memory retrieval) or related (imagination related to remembered events) to the experienced ones. According to the constructive episodic simulation hypothesis, we imagine future episodes by flexibly extracting and recombining elements of previous experiences (see chapter 2).[7] Imagination is bounded by memory, and so is creativity.[8]

DIVERSITY

What then can we do to enhance creativity? There is no magic trick, unfortunately. The content of imagination is constrained by the content of memory, and the content of memory is determined by our daily experiences and mental activities. Hence, the more our activities are related to something valuable, the more we are likely to come up with creative ideas.

Suppose you have been working on the solution to a challenging technological problem for many years. Let's also suppose that you have been paying attention to whatever else you have encountered that might be related to the problem. In that case, chances are that creative solutions to the problem could suddenly pop up in your mind when you are relaxed. Such "eureka moments" (when great insights lead to discovery or invention) often take place at unexpected times. To enhance spontaneous creativity, therefore, one must pay attention to the things that could potentially lead to valuable outcomes. This does not mean that you must narrow down your interest to one specific topic. On the contrary, breadth of experience is the key to creativity. If your experience is limited to a narrow range of topics, then the scope of your imagination will likewise be narrow, which would be harmful to creativity.

As an analogy, consider our daily activities like we're preparing ingredients for a meal. The diversity of our experiences would correspond to the diversity of our ingredients. A chef can prepare many different dishes with diverse ingredients. However, if there are limited ingredients (e.g., only vegetables), even the most versatile chef may create only a limited range of dishes. Likewise, limited interests and limited experiences will inevitably narrow the scope and diversity of ideas that can be generated by imagination. Thus, it is helpful for creativity to pay attention to and get acquainted with diverse topics because creative ideas are often generated by combining knowledge and perspectives from multiple disciplines.

Many educational institutions emphasize diversity. Interacting with people with diverse backgrounds help widen one's scope. Reading on diverse topics is also an excellent way of widening one's perspectives without spending time and energy learning from actual experiences. Innovation implicates breaking the existing frame. Because it is difficult to predict in advance which experiences will help generate creative ideas in the future, we should pay attention to diverse (but likely to be valuable) topics so that the scope of imagination widens.

OPEN-MINDEDNESS

A creative idea would be useless if not utilized. An idea will remain just an idea without an effort to implement it. Sometimes it is difficult to assess the value of a new idea. Pushing forward an idea even with uncertainty in its value would facilitate innovation. This issue holds relevance not only to personal attitudes but also to organizational and societal cultures. Anthropologists can explain a great deal about the creativity and innovation of a society by simply looking into its culture.[9] Suppose that when you propose a new idea your boss typically replies, "Don't waste time on such useless things. Focus on what you were told to do." That organization is far from being innovative. Similarly, if a society strictly prohibits ideas and expressions that stray from social norms, then that society is likely to remain static and obsolete. Group creativity is increasingly emphasized over individual creativity as technologies advance and many complex problems require multidisciplinary approaches.[10] An important factor for group creativity is open-mindedness, which facilitates unconstrained exchange of ideas.[11] For an organization and society to progress, failures and deviations must be allowed. In schools, teachers should encourage new attempts by students and be generous about failures. Open-mindedness embraces diversity, and diversity fosters creativity.

THREE Bs OF CREATIVITY

Now let's turn to more practical issues related to creativity. For the deliberate mode of creative thinking, we may consciously adopt several tactics, such as "analogy, conceptual reframing (frame shifting), finding the right question, broadening perspective, reversal, juggling induction and deduction, abduction, dissecting the problem, tinkering, and play as well as a toolkit of heuristic search methods, such as means-end analysis, hill-climbing, working backwards, and trial-and-error."[12] We cannot use such deliberate tactics to enhance spontaneous creativity. However, we may indirectly promote spontaneous creativity by promoting the activity of the default mode network. It can be helpful to relax rather than exhaust your mind when you are faced with a difficult problem. This is referred to as the three Bs of creativity.

The three Bs stand for bed, bath, and bus. When you have run out of creative ideas, it is often helpful to sleep (bed) or relax (bath) and let go of deliberate, rational thinking. Traveling to a new place (bus) and getting exposed to novel stimuli will also refresh your brain and promote chances for creative ideas. According to a recent study, novel stimuli make hippocampal-prefrontal cortical neural connections flexible.[13] The brain may seem idle when we take a rest. On the contrary, innovative ideas may appear while our minds wander freely during quiet rest and even during sleep (see chapters 1 and 3). When all deliberate attempts to solve a problem fail, try to get refreshed or relaxed and perhaps a brilliant idea will pop up spontaneously.

FLOW

Even if we try all three Bs, creative ideas won't magically appear if we are not prepared. For spontaneous creativity, we must ready our brains to generate creative ideas. In the long term, we have to pay attention to diverse subjects and widen our experiences to enhance the scope of our imagination. In the short term, we have to focus on a specific problem to facilitate spontaneous creativity to solve that particular problem. Without prior hard work and effort, simply putting yourself in the three Bs situation would not magically yield creative ideas.

In other words, conscious and deliberate efforts during non-idle states will enhance the chance of generating creative ideas during idle states. Let's assume that you need to solve an important problem (e.g., finding a novel way of proving a mathematical theorem or getting two political opponents to agree on something) within a week. In such a circumstance, focusing solely on that problem for an entire week will increase the chances of finding a solution for it.

You may experience "flow" by deeply focusing on a particular problem. Flow is the state in which someone is so completely absorbed in a challenging but doable task that they lose awareness of everything else. "Concentration is so intense that there is no attention left over to think about anything irrelevant, or to worry about problems. Self-consciousness disappears, and the sense of time becomes distorted."[14] Alternating between flow and relaxed states will increase the chances of generating creative ideas for the problem at hand. By deeply focusing on one specific issue, our neural networks will be shaped to process the information related to the problem. This will allow our imagination during the idle state to be filled with related information. Our brain will also be ready to capture a potentially creative idea when it pops up. Perhaps Archimedes went through a similar psychological process when he came up with a way to prove that a new crown made for King Hieron was not pure gold, before running naked through the streets of Syracuse shouting "Eureka!"

KOAN CONTEMPLATION ZEN

One may adopt this procedure (alternating between flow and relaxed states) when faced with a difficult problem, be it scientific, artistic, or business-related. It can be also used to obtain insights leading to spiritual enlightenment. Koan-contemplation zen is a good example of this. The goal of Buddhism is to overcome suffering by understanding how your mind works ("seeing one's original mind"). One's thoughts and feelings do not arise randomly; there are lawful dynamics behind them ("dependent arising"). You may overcome your enslavement to your thoughts and feelings by fully realizing how they work.

Koan-contemplation zen, which is widespread in east Asian countries (China, Korea, and Japan), uses focused contemplation to achieve this goal. A zen student contemplates a koan, which is a short paradoxical statement, question, or parable, such as the following:

"Two hands clap and there is a sound; what is the sound of one hand?"

"What is your original face before your mother and father were born?"

ZEN STUDENT: "What is Buddha?"
ZEN MASTER: "A shitty stick."

Koans are not logical and don't always make sense. They are intended to provoke "great doubt." They stimulate you to break out of your ordinary mental loop and expand your scope of thinking. They are a means to "give a shock to the egoistic mind shaped through conditioning and belief thinking reinforced during upbringing and conventional schooling."[15]

A breakthrough-koan (*Hua-Tou* or "'word head'") practice typically lasts three to seven days. You must devote your full attention to it day and night during this period, as described here: "A 'Hua-Tou' shines over my entire body and mind as though the bright full moon fills the sky at night. The Hua-Tou and I merge so that the distinction between the two disappears; the Hua-Tou becomes me, and I become the Hua-Tou. . . . I sink deep into the Hua-Tou rather than analyzing it as an objective. I reach to the bottom of the Hua-Tou so that there is no distance between me and the Hua-Tou, and there is no thought or feeling other than the Hua-Tou."[16]

This would correspond to a perfect flow state. Perhaps flow and relaxed states alternate during the breakthrough-koan practice so that the spontaneous creativity for generating new insights leading to spiritual enlightenment is maximized.

PERSISTENCE

We may come up with a creative idea that provides a full solution to the problem at hand. More often, however, the process of deriving the final solution is not so simple, particularly when we are faced with a complex problem. We may have to alternate periods of focused contemplation and relaxation multiple times before coming up with a sufficiently good solution to a problem. A creative idea may also turn out to be only a partial solution or it may create unexpected issues. We then have to practice the focused contemplation-relaxation cycle again (and again) to come up with a satisfactory solution. Often, therefore, you need to be persistent when you are faced with a challenging problem. "The only genuine requirements for creativity are cognitive flexibility and motivational persistence. Highly creative people will attack a problem from many different angles, enduring many false starts and dead ends, before they finally complete their quest if they manage to do so at all!"[17]

SMARTPHONES AND CREATIVITY

Imagination is fundamental to human nature. We don't need to exert special effort to unfold our imagination when we allow ourselves to ignore the outside world. Nonetheless, if we are exposed to a constant flood of stimuli, our chances to use our imagination may be seriously diminished. Many modern societies are flooded with noise and stimuli around the clock. For many of us, perhaps the single most significant source of attention-grabbing stimuli is our smartphone. People used to be engaged in internal mentation while taking a rest or in bed before falling asleep. Now, people pick up a smartphone when they have downtime and many even use it in bed before falling asleep. This enhances the activity of brain regions devoted to sensory processing while suppressing the activity of the default mode network.

Smartphone overuse also represses contemplation-related brain activity. When faced with a problem, we combine the available information in various manners and assess the outcomes to find a solution to the problem. With a smartphone, we don't have to go to such an effort; we can simply google the solution.

If continued, smartphone overuse may have long-term consequences on the brain. This can be especially problematic for users with immature brains. Some parts of the brain, such as the prefrontal cortex, undergo a maturation process until puberty. Inevitably, the long-term consequence of smartphone overuse on the brain could be more pronounced in the young than in adults. Moreover, with an immature prefrontal cortex, which plays a key role in inhibitory control (see chapter 9), it would be extremely difficult for the adolescent to refrain from smartphone overuse. For these reasons, it is important to understand the long-term consequences of smartphone overuse on the young brain.

Study results so far are worrisome. Smartphone overuse in the young is correlated with a diverse array of cognitive, emotional, and personality issues as well as neuropsychiatric disorders such as depression and attention deficit hyperactivity disorder. Overuse is also associated with weaker central executive network connectivity and a thinner cerebral cortex.[18]

In summary, the consequences of smartphone overuse on the adolescent brain could be serious. We may have to take action to alleviate it as much as we currently try to reduce rates of smoking and drinking in adolescents.

Chapter Fourteen

THE FUTURE OF INNOVATION

Is human capability for innovation a blessing? In terms of competing with other species, it definitely is. Humans' ability to innovate has allowed them to dominate other large animals. Figure 14.1 shows the weight distribution of vertebrate land animals. Ten thousand years ago, the total weight of humans comprised only 1 percent and the rest was contributed by wild animals. Now, the corresponding numbers are 32 percent and 1 percent, respectively, and the total weight of livestock (cows, sheep, horses, chickens, etc.) exceeds the combined weight of all humans and wild animals. This illustrates how dominant humans are over other large animals.

Unfortunately, the human capacity for innovation may turn out to be a curse rather than a blessing in the long run if not properly controlled. The ecosystem is being destroyed on a global scale because of human activities such as overexploitation of natural resources (see fig. 14.2), habitat destruction (see fig. 14.3), and climate change due to the overuse of fossil fuels. We may have to worry about our own extinction soon if the current trend continues.

What awaits us if we continue on our current path of innovation? Will it be a utopia or a dystopia? Predicting the future is far more difficult than looking back in time. Nonetheless, we can try to predict what will happen next based on what has happened in the past. Let's look at two opposing points of view, one pessimistic and one optimistic.

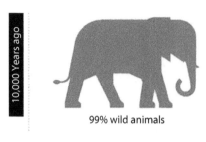

10,000 Years ago

99% wild animals

1% humans

Today

1% wild animals

32% humans

67% livestock

FIGURE 14.1. Changes in the proportion of vertebrate land animals over ten thousand years. Figure reproduced with permission from Jonathan Austern, "Save The Earth... Don't Give Birth: The Story Behind the Simplest, but Trickiest, Way to Help Save Our Endangered Planet," Infographics, accessed June 22, 2022, https://www.savetheearth.info/infographics.html.

FIGURE 14.2. Buffalo hunting in the 1800s, when guns allowed their mass hunting in North America. Figure reproduced from "File:Bison skull pile edit.jpg," Wikimedia Commons, updated May 27, 2011, https://commons.wikimedia.org/w/index.php?curid=15324296.

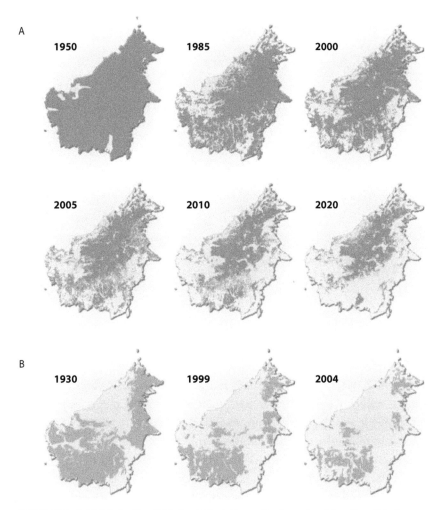

FIGURE 14.3. Habitat destruction. (A) Deforestation in Borneo. Panel reproduced from Hugo Ahlenius, "Extent of Deforestation in Borneo 1950–2005, and Projection towards 2020," GRID-Arendal, updated 2007, https://www.grida.no/resources/8324 (courtesy of Hugo Ahlenius). (B) Orangutan distribution in Borneo. Panel reproduced from Hugo Ahlenius, "Orangutan Distribution on Borneo (Indonesia, Malaysia)," updated 2006, GRID-Arendal, https://www.grida.no/resources/8326 (courtesy of Hugo Ahlenius).

PESSIMISTIC PERSPECTIVE

There have been five mass extinctions in the history of the earth. The most recent one took place sixty-five million years ago. Many animals, including all non-avian dinosaurs, went extinct. Luckily for us, mammals took this

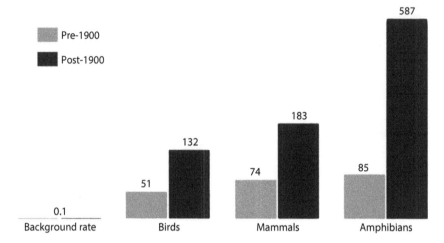

FIGURE 14.4. Recent extinction rates. Figure adapted from Hannah Ritchie and Max Roser, "Extinctions," Ourworldindata.org, accessed June 21, 2022, https://ourworldindata.org/extinctions (CC BY).

opportunity to thrive, eventually leading to the evolution of primates including humans. Now, the sixth mass extinction may be underway because of human activities. According to a study on vertebrates, the extinction rate during the past five hundred years is hundreds to thousands of times faster than an ordinary one, which is 0.1 extinction per million species per year (0.1 E/MSY), and the rate has been accelerating since 1900 (see fig. 14.4).[1] According to World Wildlife Fund's Living Planet Report 2022, the relative abundance of monitored wildlife populations decreased by 69 percent around the world between 1970 and 2018. The decline is especially steep in freshwater species populations (80 percent) and, regionally, in Latin America (94 percent).[2] The current trend of extinction, if continues, may eventually include *Home sapiens* as a victim.

The ongoing climate change is particularly concerning. All life species face the challenge of adapting to their ever-changing environment. Environmental changes may contribute to ecological diversity by forcing existing species to evolve into new species. However, if the changes are too abrupt or drastic for most living species to adapt to, they may face mass extinction.

Historically, cataclysmic events such as large-scale volcanic eruptions and asteroids induced drastic changes in the environment, killing a large portion of existing species. The most severe mass extinction, known as the *great dying*, took place 251.9 million years ago. Up to 96 percent of all marine species and around 70 percent of all land species died out. Scientists believe that massive volcanic eruptions in Siberia released immense amounts of greenhouse gases into the atmosphere over a short period. This caused abrupt global warming and a cascade of other deleterious environmental effects such as weathering (dissolving of rocks and minerals on the earth's surface), which reduced the oxygen level in the ocean. These environmental changes were too drastic for most species to adapt.[3]

The bad news is that human activities, especially the massive use of fossil fuel, appear to be inducing abrupt and drastic environmental changes that are as severe as those during past episodes of mass extinction. The fossil fuel we currently use is the outcome of a long-term carbon-trapping process. Dead plants were converted into peat instead of being degraded during the Carboniferous period (about 360 to 300 million years ago) and eventually turned into coal, oil, and gas. However, we are now reversing that process in only a few hundred years (see fig. 14.5)![4] It is not so surprising then that fossil fuel overuse induces climate change too quickly and drastically for many life species to catch up.

Why were dead plants not degraded during the Carboniferous period? According to the most popular theory, there was a sixty-million-year gap between the appearance of the first forests and wood-digesting organisms (the *evolutionary lag hypothesis*). In other words, there were no organisms

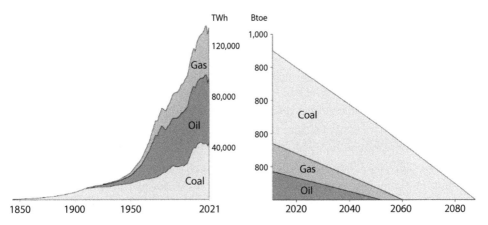

FIGURE 14.5. Fossil fuel overuse. (Left) Global fossil fuel (gas, oil, and coal) consumption, measured in terawatt-hours (TWh), between 1800 and 2019. Panel reproduced from Hannah Ritchie and Max Roser, "Energy," Ourworldindata.org, accessed June 21, 2022, https://ourworldindata.org/energy (CC BY). (Right) The prospect of future energy reserves in billion tons of oil equivalent (Btoe). The estimated endpoints for oil, gas, and coal are 2052, 2060, and 2090, respectively.

that could break down wood during that period.[5] A significant event in the history of plant evolution is the appearance of lignin, which is a class of complex organic polymers. Together with cellulose, lignin provides rigidity to land plants, allowing them to grow tall, which is advantageous for effectively harnessing the primary source of energy—sunlight. Ancient microorganisms could not break down lignin and so dead plants piled up and didn't disintegrate.[6] In a sense, lignin was like present-day plastic. It took sixty million years for lignin-digesting bacteria and fungi to evolve. Perhaps new organisms that can break down plastic will eventually evolve so that the current plastic crisis will be resolved. It just may take an extremely long time—and another sixty million years is two hundred times the entire span of existence of *Homo sapiens*.

Currently, the most significant consequence of fossil fuel overuse is global warming. According to the Intergovernmental Panel on Climate Change's (IPCC) Sixth Assessment Report, the global temperature rose approximately 1°C above pre-industrial (between 1850 and 1900) levels by 2017 (see fig. 14.6).[7] That increase in less than one hundred years is an extremely rapid change on a geological time scale; it rivals the rate of global warming during the great dying. Many organisms are dying out because they cannot keep up with such rapid environmental changes. Evolution is a slow process for most species, so if this trend of global warming continues, the rate of extinction will likely accelerate.

To strengthen the global response to the threat of climate change, 195 nations adopted the Paris Agreement in December 2015 to limit the global average temperature increase to 1.5°C above pre-industrial levels. If the participating nations faithfully abide by the agreements, i.e., reduce emissions immediately, and have CO_2 emissions reach zero by 2055, the global temperature is predicted to reach 1.5°C above pre-industrial levels around 2040 and then either reach a plateau or slightly overshoot the target.

Why 1.5°C? This is because the chance of crossing a "tipping point" is greatly increased with global warming above that level. Irreversible changes may occur and global warming may accelerate greatly once we cross that tipping point. For example, global warming decreases glacier mass, which in turn decreases glaciers' reflection of the sun's rays, resulting in further global warming. Another example is the thawing wetlands in Siberia's permafrost that release methane gas and exacerbate global warming. Above the tipping point, global warming may interact synergistically with its consequences, such as glaciers melting (see fig. 14.7), thawing of arctic wetlands,

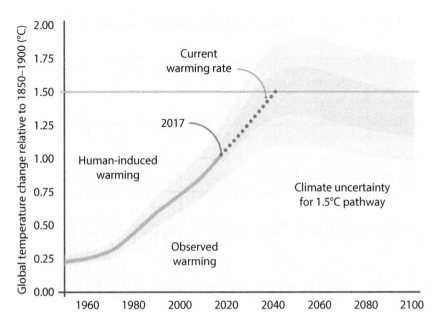

FIGURE 14.6. How close are we to 1.5°C? Figure reproduced from Myles R. Allen et al., "Framing and Context," in *Global Warming of 1.5°C*, ed. V. Masson-Delmotte et al. (Cambridge: Cambridge University Press, 2018), 82 (CC BY).

FIGURE 14.7. Flooding without rain. A record-breaking heat wave expedited glacier melting in Pakistan in April 2022, prompting a flash flood (called a glacial lake outburst flood) that collapsed the Hassanabad bridge on the Karakoram Highway. Figure reproduced from Waseem Ishaque, Rida Tanvir, and Mudassir Mukhtar, "Climate Change and Water Crises in Pakistan: Implications on Water Quality and Health Risks," *Journal of Environmental and Public Health* (November 2022):5484561 (CC).

forest destruction due to droughts and fire, and slowing of ocean circulation. Once some of these events get started, it may not be possible to reverse what human activity began.

Unfortunately, there are concerns that the actions to fulfill the pledges under the Paris Agreement are insufficient. For example, wealthy countries have not kept their pledge to provide poorer countries with one hundred billion dollars per year in climate finance. Worse, the United States, a major emitter of greenhouse gases, withdrew from the Paris Agreement in 2020 under the Trump administration. The current climate change dilemma could be a classic example of the "tragedy of the commons," which refers to a situation in which self-interested people (or organizations) with access to a common resource eventually exhaust that resource. It is a relief that the United States rejoined the Paris Agreement in 2021 under the Biden administration. Nevertheless, we cannot exclude the possibility of crossing a tipping point even if we faithfully abide by the pledges under the Paris Agreement. If we fail to do so, the chance for crossing a tipping point will increase greatly.

Global temperature changes of a few degrees may not be a big deal in the geologic time scale. The global atmospheric temperature has fluctuated substantially since the birth of life on earth. It was warmer than the present day by 5 to 10°C during the Jurassic Period (approximately 150–200 million years ago) when dinosaurs flourished, and cooler than the present day by about 6°C in the last ice age (approximately twenty thousand years ago).[8] Therefore, a change of one or two degrees in global atmospheric temperature isn't a significant event from the earth's standpoint. However, it would be a big concern for some of the planet's currently flourishing species, such as *Homo sapiens*. More than 99 percent of all species that have ever lived are now gone. The global atmospheric temperature has been stable for the last ten thousand years, during which mankind has achieved stunning success in increasing its number, expanding its territory, and building civilizations. However, if the current trend of global warming continues and crosses a tipping point, mankind may soon have to face an extremely harsh future.

OPTIMISTIC PERSPECTIVE

So far we have taken a gloomy look at the current trend of global warming. Now let's look on the bright side. First, we are aware of the potentially catastrophic consequences of global warming. Second, we have a decent scientific understanding of the human factors causing global warming.

Third, we are taking actions to mitigate it. It is, of course, difficult to predict and control climate dynamics because of its immense complexity. Nevertheless, it is encouraging that people are making an effort to understand global events and mitigate their adverse consequences.

Recognizing that certain substances, such as chlorofluorocarbons (CFCs), which were widely used as refrigerants, can damage the stratospheric ozone layer, an international treaty, the Montreal Protocol, was adopted in 1987 to protect the ozone layer by phasing out the production of ozone-depleting substances. How effective was this treaty? The estimated effect of the Montreal Protocol since 1987 amounts to the reduction of the global atmospheric temperature by 0.5–1°C by 2100.[9] This is an enormous success. The case of the Montreal Protocol delivers a powerful and encouraging message that we can stop further climate change.

Another source of optimism is the rapid pace of scientific and technical advancements. Technology grows exponentially rather than linearly. Such rapid technological advancements may provide us with the means to combat our current environmental issues. We may soon be able to develop technologies that are immensely more powerful than those now in use. We may be able to address current and future environmental challenges with the help of these technologies.

Consider how artificial intelligence is evolving. Given the current rate of scientific progress, we may soon be able to understand the key neural principles underlying human intelligence and then use this knowledge to develop artificial intelligence that outperforms human intelligence. This is already happening in some ways. For instance, we discussed how the depth of a neural network may be critical in determining the degree of abstraction it can achieve (see chapter 11). Deep learning now employs neural networks with well over a hundred layers, far exceeding the depth of the network found in the human brain.

Artificial intelligence already outperforms humans in many domains. The critical next stage is the development of a general-purpose artificial intelligence. In this regard, David Silver and colleagues at DeepMind, in their paper titled "Reward Is Enough," proposed that reward is sufficient to develop a general-purpose artificial intelligence.[10] According to their hypothesis, most, if not all, forms of intelligence, including perception, memory, planning, motor control, language, and social intelligence, emerge as a result of the overarching objective of maximizing reward. This is because maximizing reward in a complex environment requires a diverse array of abilities. Given that reinforcement learning formalizes the problem of goal-seeking intelligence, a sufficiently powerful reinforcement learning agent interacting with complex environments would develop various forms of intelligence necessary to maximize reward across multiple environments.

AlphaGo shocked the world by defeating Lee Sedol, a human Go master, in 2016. DeepMind created additional programs, such as AlphaZero and MuZero, over the ensuing years. MuZero, a powerful reinforcement learning agent, was able to master numerous board games (Go, Chess, and Shogi) as well as a number of Atari games with no prior knowledge of them. MuZero could learn to master all of these games with the single objective of maximizing reward (e.g., winning a game), despite the fact that different types of knowledge are required for different games. We may soon see the emergence of artificial general intelligence if this line of research continues.

The inventor and futurist Ray Kurzweil, highlighting the exponential nature of technological growth (the "law of accelerating returns"), predicted that "the technological singularity" (the point at which technological growth becomes uncontrollable and irreversible) would occur in 2045.[11] Once we reach the singularity, artificial intelligence will surpass human intelligence and continue to improve at an ever-accelerating rate without human intervention, triggering a self-improvement runaway reaction. In 1966, the mathematician Irving Good predicted an "intelligence explosion" and the emergence of artificial superintelligence as follows: "Let an ultraintelligent machine be defined as a machine that can far surpass all the intellectual activities of any man however clever. Since the design of machines is one of these intellectual activities, an ultraintelligent machine could design even better machines; there would then unquestionably be an 'intelligence explosion,' and the intelligence of man would be far left behind. Thus the first ultraintelligent machine is the *last* invention that man need ever make, provided that the machine is docile enough to tell us how to keep it under control."[12]

One hopeful outcome of this intelligence explosion is the development of new and powerful technologies capable of halting or even reversing current environmental problems. This may not appear plausible from the standpoint of contemporary technology. However, once technological advances reach a tipping point, progress may far outpace what we can currently imagine.

Can the pace of technological progress continue to speed up indefinitely? Isn't there a point at which humans are unable to think fast enough to keep up?

For unenhanced humans, clearly so. But what would 1,000 scientists, each 1,000 times more intelligent than human scientists today, and each operating 1,000 times faster than contemporary humans (because the information processing in their primarily nonbiological brains is faster) accomplish?. . . When scientists become a million times more intelligent and operate a million times faster, an hour would result in a century of progress (in today's terms).[13]

Kurzweil predicted a number of events that would occur following the singularity, including the ability of nanobots to reverse the pollution caused by earlier industrialization.[14] I hope anticipating technology that can stop current global warming is not naive optimism. Exponential knowledge growth, which includes advances not only in science and technology, but also in our understanding of human nature and society, as well as optimizing strategies for organizing and mobilizing social resources, may provide us with effective tools to address the current environmental challenges.

Linear and exponential growth appear comparable in the early stages. Their differences become apparent only when the growth reaches a certain threshold (the knee of the exponential curve; see fig. 14.8). From this perspective, it appears that we are currently living in an era in which the exponential growth of technology is beginning to accelerate past a critical point.

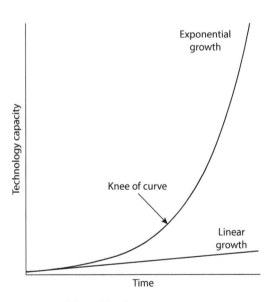

FIGURE 14.8. Linear versus exponential growth functions.

Let me illustrate this idea with a few examples derived from my personal experiences. In 1981, as a biology student, I took a molecular genetics course. I recall the awe I felt when I first learned about the molecular processes underlying the central dogma (the transfer of information from DNA to messenger RNA to protein), which is a fundamental principle underlying all life forms on earth. The potential of gene cloning (or recombinant DNA) technology was mentioned only briefly in the textbook's last chapter. Gene cloning is the process of making multiple copies of a gene. Using restriction enzymes (enzymes that cut DNA at certain nucleotide sequences), a specific DNA sequence is cut, inserted into a vector (a small piece of DNA that can replicate independently of the host genome), and then introduced into a host cell. This marked the beginning of genetic engineering. Since then, genetic engineering technologies have advanced rapidly. CRISPR-CAS, a relatively new technology, in particular, enables precise editing of DNA sequences in living organisms. These genetic engineering technologies are now being used in conjunction with other technologies, such as big data science and artificial intelligence, to design and build new biological parts, devices, and systems that do not exist in nature. One could, for example, try to design and create new microorganisms that degrade plastic. This is an emerging field known as synthetic biology. Scientists confirmed that DNA is the genetic substance in 1944.[15] Less than a hundred years later, scientists have identified the genetic code (a triplet of nucleotides specifying one amino acid), molecular processes underlying the central dogma, and the entire human genome sequence, and they can now design and manufacture new life forms.

Rapid advances in computer technology are another example. Another course I took in 1981 was computer programming. Back then, we had to write codes on squared paper and submit them to a punch-card operator. After a few days, we would receive a thick bundle of punch cards, which we then brought to a computer center for the main computer to read and generate the output. The main computer read the data from the punch cards (based on the absence or presence of a hole at a specific location) and generated the final output. Even a very short program required a thick stack of punch cards.

I first encountered a personal computer when I went to the United States for graduate school in 1986. It was a whole new experience for me—my inputs immediately produced results on the computer screen. I could even write and run a program without using punch cards! The hard disk of the computer I was using at the time—I think it was an IBM XT—had to be backed up one day because it was almost full. The total disk space of the computer was 10 MB. I used about twenty or so floppy disks (525 KB each) to back up the disk. Now, about thirty-five years later, my smart-phone has 256 GB of disk space, which is 25,600 times more than that computer from the 1980s. The speed and capacity of computers have been doubling approximately every two years for the past few decades, accord-ing to Moore's law. This means that today's computers are about 100 times more powerful than those 14 years ago, 10,000 times more powerful than those 27 years ago, and 1 million times more powerful than those 40 years ago. As such, the advancement of computer technology has been visibly exponential for the past few decades.

Artificial intelligence (AI) is another field that is rapidly progressing. Since the deep learning revolution, many powerful AI programs have emerged that far outperform humans in various domains. Even in fields such as art, music, and literature, which were once thought to require special human abilities, powerful AI programs are now available. They can create music, drawings, videos, or even poems and novels. The following is a poem chatGPT wrote after I requested it to "write a poem on human creativity."

> Creativity flows like a river wild,
> A force that cannot be tamed or reconciled.
> It bursts forth from the depths of the mind,
> A spark that sets the imagination aligned.
>
> It dances in the brushstrokes of a painter,
> It sings in the notes of a composer's gainer.
> It shapes the words of a poet's pen,
> And the designs of an engineer's den.
>
> It builds the bridges that span the divide,
> And the skyscrapers that reach for the sky.
> It brings to life stories that touch the soul,
> And the inventions that make us whole.
>
> Human creativity is a wonder to behold,
> A reflection of the world, both young and old.
> It is the light that guides us through the dark,
> A reminder of the beauty in our hearts.

ChatGPT wrote this poem in a matter of seconds. It'll probably take me a month or more to write a poem as good as this one.

If powerful AI technologies continue to advance exponentially and are combined with technological breakthroughs in other areas, they may become effective in combating the environmental issues we are currently facing. DeepMind was able to save 40 percent of total cooling costs by applying AI technologies to the control of Google data centers.[16] This is a significant improvement given that Google data centers were already managed in a highly sophisticated manner prior to the application of DeepMind's AI. This example shows how AI can be used to improve energy efficiency and minimize emissions.

CROSSROADS

We might be approaching a crossroads where, on the one hand, technological advancements begin to take off and, on the other hand, excessive exploitation of natural resources threatens the continued existence of our own species. The COVID-19 pandemic might be evidence of both of these trends. The pandemic is the result of the spread of the SARS-CoV-2 virus, which is believed to have originated in bats before spreading to humans. As people continue to invade wildlife habitats, the risk of zoonotic disease (animal-borne infectious disease for humans) will inevitably increase. There is always the possibility of a new zoonotic disease, such as COVID-19, that is both highly infectious and fatal. However, advances in science and technology have provided us with the means to deal with COVID-19. Some countries were able to keep their case numbers and fatalities relatively low by implementing early lockdown procedures, widespread testing, contact tracing, isolation, and effective communications with citizens. Various technologies, such as disease diagnosis, the internet, and smartphones, played roles in this process. More significantly, we were able to develop effective vaccines in less than a year after the COVID-19 outbreak, which is unprecedented in human history. This was made possible by the vast corpus of technological and scientific advancements humanity has amassed over centuries in a range of disciplines, including biology, chemistry, and engineering.

The genus *Homo* is estimated to have appeared approximately three million years ago. Many *Homo* species have since gone extinct and currently only one, *Homo sapiens*, is alive. Neanderthals and Denisovans existed for about two to three hundred thousand years and *Homo erectus* for about two million years.[17] And *Homo sapiens*? Approximately three hundred thousand years so far. How much longer can we survive? We cannot anticipate a bright future if the current trend of global environmental changes continues. We cannot exclude the possibility that *Homo sapiens*—and by extension, the entire genus *Homo*—will go extinct in the not-too-distant future. Alternatively, the rapid advances in science and technology may offer us effective ways to battle the consequences of global environmental changes.

Can humans use their capability for innovation to overcome the crisis of global environmental change and sustain civilizations? Or will they run out of time in dealing with the impending environmental disaster? Our journey may lead us where scientific and technological advancements enable humans to overcome current and future challenges and transcend biological constraints (utopia). On the contrary, our path may lead to the bleak consequences of a new mass extinction; we may end up facing extremely harsh conditions, including food scarcity, that will demolish the infrastructure of the civilizations humanity has built throughout history (dystopia). The remainder of the twenty-first century may prove crucial for humanity's future.

EPILOGUE

In this book, we explored the neural underpinning of the human capacity for innovation, with a specific focus on the neural mechanisms involved in imagination and abstraction. We began our journey with the finding that the hippocampus, a brain structure well known for its role in learning and memory, is also involved in imagination (part 1). In other words, memory and imagination are supported by the same brain structure, which may explain why our memories are malleable and why we sometimes form false memories of events that did not occur.

We then explored hippocampal neural circuit processes underlying imagination (part 2). The CA3 region of the hippocampus, whose neurons are connected to each other by massive recurrent projections and hence prone to self-excitation, generates both experienced and unexperienced (novel) activity sequences during sleep and idle states. CA3 neurons are interconnected by a huge number of individually weak synapses, making them ideal for generating variable, rather than fixed, activity sequences. The CA3 network appears to have gained a degree of randomness during evolution.

We went on to examine the simulation-selection model, which posits that the CA3-CA1 neural network simulates and reinforces neural representations for high-value events and actions in preparation for the future rather than simply recalling what has already happened. We also explored the hypothesis that the hippocampus's simulation-selection function has evolved in land-navigating mammals, but not in birds, because of the necessity to choose optimal trajectories between two arbitrary locations.

Studies indicate that similar hippocampal neural processes underpin memory and imagination in humans and other mammals. The simulation-selection model posits that all land-navigating mammals have similar hippocampal neural processes that support imagination. In other words, the ability to imagine appears to be shared by all mammals. What then makes humans so innovative? We explored the neural underpinning of high-level abstraction in humans with the premise that humans are particularly innovative because they can imagine freely using high-level abstract concepts (part 3). The prefrontal cortex and the precuneus were examined as potential brain regions critical for high-level abstraction in humans. We also considered the possibility that the depth (the number of layers) of a neural network might determine the level of abstraction.

How much do we know about the neural underpinning of human innovation? Have we identified the critical neural processes that underpin imagination and high-level abstraction? Unfortunately, not yet. We are only beginning to understand the neural underpinning of this great human mental faculty. Furthermore, what we covered in this book only scratches the surface of the subject. First, in terms of the neural basis of imagination, the hippocampus does not function independently but interacts with a large number of brain regions during imagination. In this book, we only looked at hippocampal neural processes. Second, the neural basis of high-level abstraction is less well understood compared to that of imagination. As a result, our discussions on this subject (part 3) are necessarily limited. Third, as we discussed in chapter 13, imagination alone is insufficient for creativity and innovation. An evaluative process that constrains ideas to meet a specific need should follow the generation of ideas. We concentrated on the former (spontaneous creativity) while ignoring the latter (deliberate creativity). Please be aware of these constraints.

Part 4 covered some topics related to but distinct from the book's main thesis (the brain foundation of imagination and high-level abstraction), such as language and how to become creative. In the final chapter, we contrasted two different perspectives on the consequences of innovation, namely the sustainability problem and intelligence explosion. It is noteworthy that we have made significant progress in artificial intelligence despite our limited understanding of the neural basis of human innovation. Artificial neural networks are superior to the human brain in terms of signal processing speed (GHz versus less than 100 Hz), network depth (more than a hundred layers versus less than twenty layers; see chapter 11), evolution speed (days, months, or years for a new version of an artificial neural network versus multiple generations for natural evolution), and technology spread (instantaneous duplication versus an inability to duplicate). These factors enabled artificial neural networks to outperform human intelligence in many domains. We may be able to build much more powerful artificial neural networks in the near future as we better understand the neural basis of human innovation and apply this knowledge to the construction of next-generation artificial neural networks. Then, as Kurzweil predicted, "the singularity" may be near.[1]

As elaborated in chapter 14, we may be on the verge of a tipping point that will lead to either a utopia or a dystopia. Our capacity for innovation is a double-edged sword. We need to make full use of it to resolve the crisis resulting from its misuse rather than facilitate it. We will most likely need to take a multifaceted approach. We need technological breakthroughs that can effectively mitigate the effects of climate change. At the same time, we need to understand ourselves better in order to foster global collaboration. We need to make the best use of all available social resources in order to address the ongoing environmental challenges.

DENTATE GYRUS

Of the hippocampal trisynaptic circuit, we focused on the CA3 and CA1 regions. This is because those regions are thought to be directly related to the main topic of this book—the neural basis of imagination. What then does the dentate gyrus do? Currently, the most widespread theory is that it performs pattern separation, which enhances the memory storage capacity of the CA3 neural network. As an alternative account, together with my long-term colleague, Jong Won Lee, I proposed that the main function of the dentate gyrus is to bind together diverse sources of incoming sensory inputs to set a spatial context.[1] Here, I will briefly explain each theory and summarize theoretical issues associated with the pattern separation theory.

PATTERN SEPARATION

According to this theory, the major function of the dentate gyrus is to separate similar input patterns into distinct ones so that the CA3 neural network can store many input patterns with minimal interference. The theory is based on the dentate gyrus having a particularly large number of neurons. Earlier, in his theory on the cerebellum, David Marr proposed that overlapping activity patterns of pontine mossy fibers are transformed into nonoverlapping ones by divergent projections to granule cells (expansion recoding).[2] In the rat hippocampus, the dentate gyrus has about one million excitatory neurons (granule cells), which are more than CA3 and CA1

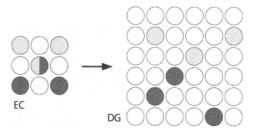

FIGURE A1.1. Pattern separation by expansion recoding (EC: entorhinal cortex; DG: dentate gyrus).

excitatory neurons (pyramidal cells) combined. Dentate gyrus granule cells receive main excitatory inputs from about 0.2 million cells in layer 2 of the entorhinal cortex.[3] Thus, as in the cerebellum, the projection from a smaller neural network (entorhinal cortex layer 2) to a larger network (dentate gyrus) may reduce activity pattern overlaps in the hippocampus (see fig. A1.1).[4] This way, according to the pattern separation theory, the dentate gyrus helps CA3 store many patterns with minimal interference.

BINDING

The entorhinal cortex has two divisions: medial and lateral. The medial entorhinal cortex, where grid cells are found (fig. 8.2), is believed to carry a general metric representation of space, whereas the lateral entorhinal cortex is believed to carry information about the specifics of an environment. More generally, the medial and lateral entorhinal cortex may carry abstract structural knowledge and individual sensory experiences, respectively (see chapter 8). Anatomically, these two input streams converge in the dentate gyrus and CA3, suggesting that the dentate gyrus and CA3 may play an important role in binding together different types of inputs to the hippocampus (see fig. 8.4).[5] In the spatial domain, a general spatial representation (from the medial entorhinal cortex) and information about the specifics of an environment (from the lateral entorhinal cortex) converge in the dentate gyrus and CA3. These signals, when combined, would be sufficient to set a spatial context for a given environment (knowing where one is).

Relative contributions of the dentate gyrus versus CA3 to binding incoming sensory inputs are unclear. Nevertheless, inputs from the entorhinal cortex (medial and lateral combined) constitute about two-thirds of all excitatory synaptic inputs to dentate gyrus granule cells, whereas they comprise less than one-quarter of the inputs in CA3 pyramidal cells (see fig. 4.2). This anatomical organization suggests that entorhinal cortical inputs have a relatively important role in dentate gyrus functioning. Moreover, a growing body of experimental evidence points to the role of the dentate gyrus in sensory binding and spatial context setting.[6] These findings indicate that the dentate gyrus, alone or together with CA3, is involved in sensory binding and spatial context setting. In fact, this proposal is not entirely new. Raymond Kesner has proposed that the dentate gyrus plays multiple roles including pattern separation and "conjunctive encoding," and Sen Cheng has proposed "context reset" as the primary function of the dentate gyrus.[7]

THEORETICAL ISSUES WITH PATTERN SEPARATION

Pattern separation and binding are not mutually exclusive. The dentate gyrus may bind together incoming sensory inputs from the entorhinal cortex and, in this process, separate overlapping patterns to enhance a "sparseness of representation." However, there are reasons to suspect that pattern separation may not capture the essence of the dentate gyrus's contribution to hippocampal functioning. Detailed arguments against pattern separation and for binding can be found in our paper.[8] Here, we will briefly examine only the theoretical issues associated with the pattern separation theory.

First, from an evolutionary standpoint, it is doubtful that the dentate gyrus has evolved primarily for pattern separation. Several lines of evidence indicate that it evolved de novo in mammals; birds do not have a homologous structure.[9] As we examined in chapter 7, some birds (ones that cache food) have superb spatial memory capacity. They store food in thousands of different areas and successfully retrieve most of it many months later based on spatial memory. This means that these birds found a way to represent many memories with little interference without a dentate gyrus. So why mammals have evolved such a huge structure (dentate gyrus) when there is a simpler solution to enhance memory storage capacity?[10]

Second, the pattern separation theory concerns the separation of static patterns, but it is now clear that CA3 represents sequence memories as well (such as spatial trajectories; see chapter 3). For static patterns, it is conceivable that projections from a smaller network to a larger network may reduce overlaps among patterns by expansion recoding (see fig. A1.1). However, the effect of expansion recoding on the separation of sequences is not straightforward because a sequence of neural activity depends not only on incoming inputs but also on the internal dynamics of a neural network.

Third, it is difficult to reconcile pattern separation and synaptic plasticity (a change in neural connection strength) found in the dentate gyrus.[11] Synaptic plasticity in the dentate gyrus suggests that it plays a role in encoding new memories. Scientists believe that a neural network stores memories in an overlapping and distributed manner by changing connection strengths among neurons (see figs. 4.3 and 8.1). Then those synapses commonly activated by similar inputs will be repeatedly strengthened (this is a neural basis of generalization; see chapter 8). Consequently, for sufficiently similar inputs, a neural network will perform pattern completion (make similar inputs more similar) rather than pattern separation (make similar inputs more different). Assuming that the dentate gyrus stores memories according to this general scheme, synaptic plasticity would be detrimental to the proposed role of the dentate gyrus in separating similar input patterns into distinct ones.

To summarize, other than the idea of expansion recoding, theoretical grounds for the pattern separation theory are tenuous. Empirical studies have also failed to provide clear evidence for pattern separation. For example, although numerous studies showed that dentate gyrus manipulations impair an animal's discrimination between two similar stimuli (such as two close spatial locations), such behavioral deficits may be caused by numerous factors. Impaired behavioral discrimination may be caused by impaired pattern separation; however, it may be caused by deficits in other neural processes, such as an impaired binding of incoming sensory inputs. Please see our paper already referenced in the Notes for more detailed discussions of this matter.

VALUE-CODING NEURONS

Value-coding neurons are found in widespread areas of the brain. There are different types of value (such as value of an object, value of a particular action, or value of the current situation) and different neurons encode them. A value-coding neuron changes its spiking activity according to value. In other words, we can gain information about value by observing a neuron's spiking activity. Some value-coding neurons increase their firing rates as the value increases, and some decrease their activity as the value increases.[1] Some neurons are also responsive to a particular range of value.[2] In all of these cases, the spiking activity of a value-coding neuron provides information about value.

Here, I will explain how scientists find value-coding neurons with a specific example—a CA1 neuron recorded from a rat hippocampus that increases spiking activity as value increases. My team trained a thirsty rat to obtain a water reward by visiting two target locations (upper left and upper right corners in a modified T-maze; see fig. A2.1). In each trial, starting from the central stem, the rat had to visit one of the two targets and come back to the center using the lateral alley (gray arrows indicate the rat's movement directions). Each target delivered a small drop of water probabilistically. Thus, the rat drank water in some trials but could not in others. Moreover, the probabilities of water delivery at the two targets changed over time without any sensory cue. For example, the probabilities for the left and right targets unpredictably changed from 21 percent and 63 percent to 72 and 12 percent.

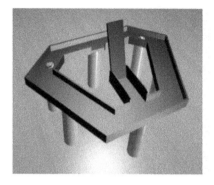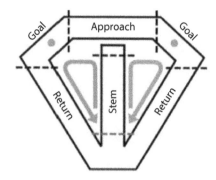

FIGURE A2.1. A modified T-maze. Arrows indicate movement directions of the rat. Gray circles are the two water-delivery locations. Right panel reproduced from Hyunjung Lee et al., "Hippocampal Neural Correlates for Values of Experienced Events," *Journal of Neuroscience* 32, no. 43 (October 2012): 15054 (CC BY-NC-SA).

This dynamic foraging task is used to emulate decision-making in a dynamic and uncertain environment, where precisely keeping track of true values is the biggest challenge. The rat's challenge is to keep track of the water-delivery probabilities of the two targets. This can be achieved by estimating the probabilities based on the history of past choices and their outcomes. Let's assume that the rat chose each target five times over ten trials. Let's also assume that the rat drank water in four of the five visits at the left target and only one of the five visits at the right target. The rat probably estimates the water-delivery probability of the left target (i.e., the value of the left target) to be higher than that of the right. We can then predict that the rat will likely choose the left target in the next trial.

Now let's consider another situation. Let's assume that the rat drank water in three out of five visits at each target over ten trials. Let's also assume that the sequence of choice outcomes was o-o-o-x-x ("o" and "x" indicate water delivery and no delivery, respectively) at the left target and x-x-o-o-o at the right target. Which target would you consider to be of higher value? The right target, presumably. This is because we give more weight to recent than remote experiences when evaluating them. This is reasonable because we live in an ever-changing world. The two sequences may be because the true value of the right target increased while that of the left target decreased. The reinforcement learning theory formalizes these processes—valuation of experienced outcomes with more weight on recent experiences.

The reinforcement learning theory also formalizes the process of making a choice while taking into account the tradeoff between exploitation and exploration. How would you distribute your choices to maximize the amount of reward in a dynamic and uncertain environment? You must choose the target with the highest estimated value over the other targets, of course. However, you also have to choose the other targets occasionally because values may change dynamically over time. A previously low-value target may become a high-value target. Also, a newly available target may turn out to be a very high-value target. This well-known problem in reinforcement learning is called the exploitation-exploration tradeoff. Should you exploit the highest-value target or explore your options by choosing a low-value target or a new one to gain information?[3]

As elaborated in chapter 5, the reinforcement learning theory is widely used to investigate the neural basis of value-based decision-making. Indeed, the rat's choice behavior in this task is well predicted by a simple reinforcement learning algorithm. Using this algorithm (the Q-learning model, to be specific) we estimated values for the left and right targets in each trial that varied between 0 and 1 (estimated water-delivery probabilities).[4] We implanted microelectrodes in CA1 and recorded spiking activities of single neurons while the rat was performing the task (about 150 trials). To our surprise, we found a substantial fraction of CA1 neurons whose activity was highly correlated with the value of either the left or right target.[5] In other words, many CA1 neurons conveyed information about the value of the left or right target.

Figure A2.2 shows an example CA1 neuron that conveyed information about the value of the left target. The modified T-maze was linearized for simplicity. The left end of the plot corresponds to the central stem and the right end corresponds to the lateral alleys (the left and right alleys were combined in the plot). The top plot shows individual spikes (dots) of this neuron. As you can see, most spikes were emitted when the rat was at the lower part of the central stem (left end of the linearized plot). Hence, this is a typical hippocampal place cell. However, if we divide all the trials (about 150 total) into four groups according to the estimated value of the left target (0~0.25, 0.25~0.5, 0.5~0.75, and 0.75~1), it is evident that this place cell emits more spikes as the value of the left target increases.

The bottom plot shows the firing rate (the number of spikes divided by occupancy time) of this neuron on the linearized maze. The firing rate of this neuron is the highest in the lower part of the central alley, and it increases as

VALUE-CODING NEURONS

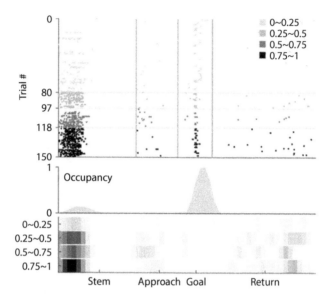

FIGURE A2.2. A value-coding CA1 neuron. The maze shown in figure A2.1 was linearized. (Top) Trial-by-trial spiking activity. Each line is one trial, and each dot represents a spike. The trials were divided into four groups based on the value of the left target estimated using a Q-learning model. (Middle) Rat's occupancy time. The rat stayed longer at the goal sites than at the other locations. (Bottom) Spatial firing rate (spikes/occupancy) on the linearized maze (darkness indicates high spatial firing). Trials were grouped into four as in the top plot. Figure adapted from Hyunjung Lee et al. "Hippocampal Neural Correlates for Values of Experienced Events," *Journal of Neuroscience* 32, no. 43 (October 2012): 15060 (CC BY-NC-SA).

the value of the left target increases (high firing rate is indicated by darkness). Thus, this CA1 neuron conveyed both spatial and value information. It is a place cell and also a value-coding neuron. As illustrated in this example, a neuron whose activity is significantly correlated with value (according to a standard statistical criterion, of course), hence conveying value information, is considered a value-coding neuron.

NOTES

INTRODUCTION

1. Smithsonian, "Numbers of Insects (Species and Individuals)," Buginfo, Information Sheet Number 18, 1996, https://www.si.edu/spotlight/buginfo/bugnos; Entomological Society of America, "Frequently Asked Questions on Entomology," updated July 26, 2010, https://www.entsoc.org/resources/faq/; Royal Entomological Society, " Facts and Figures," Understanding Insects, n.d., https://www.royensoc.co.uk/facts-and-figures.
2. Nigel E. Stork, "How Many Species of Insects and Other Terrestrial Arthropods Are There on Earth?," *Annual Review of Entomology* 63 (January 2018): 32, 37.

1. HIPPOCAMPUS: FROM MEMORY TO IMAGINATION

1. Darryl Bruce, "Fifty Years Since Lashley's 'In Search of the Engram': Refutations and Conjectures," *Journal of the History of the Neurosciences* 10, no. 3 (2001): 315.
2. Larry R. Squire and Pablo Alvarez, "Retrograde Amnesia and Memory Consolidation: A Neurobiological Perspective," *Current Opinions in Neurobiology* 5, no. 2 (April 1995): 172.
3. James L. McClelland, Bruce L. McNaughton, and Randall C. O'Reilly, "Why There Are Complementary Learning Systems in the Hippocampus and Neocortex: Insights from the Successes and Failures of Connectionist Models of Learning and Memory," *Psychology Review* 102, no. 3 (July 1995): 424–25, 440, 447, 453.
4. Lynn Nadel et al., "Multiple Trace Theory of Human Memory: Computational, Neuroimaging, and Neuropsychological Results," *Hippocampus* 10, no. 4 (2000): 358–65.
5. According to this theory, seemingly preserved remote episodic memories in people with hippocampal injury may, in fact, be semantic memories.
6. William B. Scoville and Brenda Milner, "Loss of Recent Memory After Bilateral Hippocampal Lesions," *Journal of Neurology and Neurosurgical Psychiatry* 20, no. 1 (February 1957): 16–67; Suzanne Corkin, "What's New with the Amnesic Patient

H. M.?" *Nature Reviews Neuroscience* 3, no. 2 (February 2002): 15; Larry R. Squire, "The Legacy of Patient H. M. for Neuroscience," *Neuron* 61, no. 1 (January 2019): 6.

7. Demis Hassabis et al., "Patients with Hippocampal Amnesia Cannot Imagine New Experiences," *Proceedings of the National Academy of Sciences of the United States* 104, no. 5 (January 2007): 1726–31.
8. Donna R. Addis, Alana T. Wong, and Daniel L. Schacter, "Remembering the Past and Imagining the Future: Common and Distinct Neural Substrates During Event Construction and Elaboration," *Neuropsychologia* 45, no. 7 (April 2007): 1363–77; Karl K. Szpunar, Jason M. Watson, and Kathleen B. McDermott, "Neural Substrates of Envisioning the Future," *Proceedings of the National Academy of Sciences of the United States* 104, no. 2 (January 2007): 642–7.
9. Randy L. Buckner, Jessica R. Andrews-Hanna, and Daniel L. Schacter, "The Brain's Default Network: Anatomy, Function, and Relevance to Disease," *Annals of the New York Academy of Science* 1124 (March 2008): 2–3.
10. Marcus E. Raichle et al., "A Default Mode of Brain Function," *Proceedings of the National Academy of Sciences of the United States* 98, no. 2 (January 2001): 676, 682.
11. Buckner, Andrews-Hanna, and Schacter, "The Brain's Default Network," 20–24; Reece P. Roberts and Donna R. Addis, "A Common Mode of Processing Governing Divergent Thinking and Future Imagination," in *The Cambridge Handbook of the Neuroscience of Creativity*, ed. Rex E. Jung and Oshin Vartanian (Cambridge: Cambridge University Press, 2018), 213–15.
12. News Staff, "Breakthrough of the Year: The Runners-Up," *Science* 318, no. 5858 (December 2007): 1848–49a.

2. FALSE MEMORY

1. Henry L. Roediger and Kathleen B. McDermott, "Creating False Memories: Remembering Words Not Presented in Lists," *Journal of Experimental Psychology: Learning, Memory, and Cognition* 21, no. 4 (1995): 803–14.
2. Michael Craig, "Memory and Forgetting," in *Encyclopedia of Behavioral Neuroscience*, ed. Sergio Della Sala et al. (Amsterdam: Elsevier, 2021), 425–31.
3. Sanjida O'Connell, "The Perils of Relying on Memory in Court." *Telegraph*, December 15, 2008, https://www.telegraph.co.uk/technology/3778272/The-perils-of -relying-on-memory-in-court.html.
4. Elizabeth F. Loftus and Katherine Ketcham, "Truth or Invention: Exploring the Repressed Memory Syndrome; Excerpt from 'The Myth of Repressed Memory,'" *Cosmopolitan*, April 1995, https://staff.washington.edu/eloftus/Articles/Cosmo.html; Lauren Slater, *Opening Skinner's Box: Great Psychological Experiments of the Twentieth Century* (New York: Norton, 2005), 181–203; Buckner F. Melton, Jr., "George Franklin Trial: 1990–91," Encyclopedia.com, accessed June 28, 2022, https://www.encyclopedia .com/law/law-magazines/george-franklin-trial-1990-91; Stephanie Denzel, "George Franklin," The National Registry of Exonerations, updated May 2, 2022 https://www .law.umich.edu/special/exoneration/Pages/casedetail.aspx?caseid=3221.
5. "Rodney Halbower," Wikipedia, updated April 27, 2022, https://en.wikipedia.org/wiki /Rodney_Halbower.
6. Lawrence Wright, "Remembering Satan—Part II. What Was Going On in Thurston County?," *New Yorker*, May 16, 1993, https://www.newyorker.com/magazine/1993 /05/24/remembering-satan-part-ii; Mark L. Howe and Lauren M. Knott, "The Fallibility

of Memory in Judicial Processes: Lessons from the Past and Their Modern Conse-quences," *Memory* 23, no. 5 (2015): 636–47; "Thurston County Ritual Abuse Case," Wikipedia, updated December 9, 2021, https://en.wikipedia.org/wiki/Thurston _County_ritual_abuse_case.

7. Richard J. Ofshe, "Inadvertent Hypnosis During Interrogation: False Confession Due to Dissociative State; Mis-identified Multiple Personality and the Satanic Cult Hypothesis," *International Journal of Clinical and Experimental Hypnosis* 40, no. 3 (1992): 152.

8. Wright, "Remembering Satan"; Ofshe, "Inadvertent Hypnosis During Interrogation," 125–56.

9. Karen A. Olio and William F. Cornell, "The Facade of Scientific Documentation: A Case Study of Richard Ofshe's Analysis of the Paul Ingram case," *Psychology, Public Policy, and Law* 4, no. 4 (1998): 1194–95.

10. Elizabeth F. Loftus, "Planting Misinformation in the Human Mind: A 30-Year Inves-tigation of the Malleability of Memory," *Learning & Memory* 12, no. 4 (2005): 361–66.

11. Elizabeth F. Loftus and Jacqueline E. Pickrell, "The Formation of False Memories," *Psychiatric Annals* 25, no. 12 (December 1995): 720–25.

12. Kimberley A. Wade et al., "A Picture Is Worth a Thousand Lies: Using False Photo-graphs to Create False Childhood Memories," *Psychonomic Bulletin & Review* 9, no. 3 (September 2002): 597–603.

13. Daniel L. Schacter, "Constructive Memory: Past and Future," *Dialogues in Clinical Neuroscience* 14, no. 1 (March 2012): 8.

14. Schacter, "Constructive Memory," 11.

3. PLACE CELLS AND HIPPOCAMPAL REPLAY

1. T. V. Bliss and A. R. Gardner-Medwin, "Long-Lasting Potentiation of Synaptic Trans-mission in the Dentate Area of the Unanaesthetized Rabbit Following Stimulation of the Perforant Path," *Journal of Physiology* 232, no. 2 (July 1973): 357–74; T. V. Bliss and T. Lomo, "Long-Lasting Potentiation of Synaptic Transmission in the Dentate Area of the Anaesthetized Rabbit Following Stimulation of the Perforant Path," *Journal of Physiology* 232, no. 2 (July 1973): 331–56.

2. Steve Ramirez et al., "Creating a False Memory in the Hippocampus," *Science* 341, no. 6144 (July 2013): 387–91.

3. John O'Keefe and Jonathan Dostrovsky, "The Hippocampus as a Spatial Map: Prelimi-nary Evidence from Unit Activity in the Freely-Moving Rat," *Brain Research* 34, no. 1 (November 1971): 171–75.

4. John O'Keefe and Lynn Nadel, *The Hippocampus as a Cognitive Map* (Oxford: Clarendon, 1978), 217–30.

5. Arne D. Ekstrom et al., "Cellular Networks Underlying Human Spatial Navigation," *Nature* 425, no. 6954 (September 2003): 184–88.

6. Nobelförsamlingen, "Press release. The 2014 Nobel Prize in Physiology or Medicine. 2014," October 6, 2014, https://www.nobelprize.org/prizes/medicine/2014/press -release/.

7. For example, William B. Levy, "A Sequence Predicting CA3 Is a Flexible Associator That Learns and Uses Context to Solve Hippocampal-Like Tasks,"*Hippocampus* 6, no. 6 (1996): 579–90.

8. Gyorgy Buzsaki, "Hippocampal Sharp Wave-Ripple: A Cognitive Biomarker for Episodic Memory and Planning," *Hippocampus* 25, no. 10 (October 2015): 1073.

9. Larry R. Squire and Pablo Alvarez, "Retrograde Amnesia and Memory Consolidation: A Neurobiological Perspective," *Current Opinions in Neurobiology* 5, no. 2 (April 1995): 171–72.

10. Kenway Louie and Matthew A. Wilson, "Temporally Structured Replay of Awake Hippocampal Ensemble Activity During Rapid Eye Movement Sleep," *Neuron* 29, no. 1 (January 2001): 145–56.

11. Albert K. Lee and Matthew A. Wilson, "Memory of Sequential Experience in the Hippocampus During Slow Wave Sleep," *Neuron* 36, no. 6 (December 2002): 1183–94.

12. An earlier study by Skaggs and McNaughton also showed that the temporal order of activity between two place cells during sleep reflects that during spatial exploration before sleep. William E. Skaggs and Bruce L. McNaughton, "Replay of Neuronal Firing Sequences in Rat Hippocampus During Sleep Following Spatial Experience," *Science* 271, no. 5257 (March 1996): 1870–73.

13. David J. Foster and Matthew A. Wilson, "Reverse Replay of Behavioural Sequences in Hippocampal Place Cells During the Awake State," *Nature* 440, no. 7084 (March 2006): 680–83; Kamran Diba and Gyorgy Buzsaki, "Forward and Reverse Hippocampal Place-Cell Sequences During Ripples," *Nature Neuroscience* 10, no. 10 (October 2007): 1241–42.

14. Yvonne Y. Chen et al., "Stability of Ripple Events During Task Engagement in Human Hippocampus," *Cell Reports* 35, no. 13 (2021): 109304; Anli A. Liu et al., "A Consensus Statement on Detection of Hippocampal Sharp Wave Ripples and Differentiation from Other Fast Oscillations," *Nature Communications* 13 (2022): 6000.

15. Anoopum S. Gupta et al., "Hippocampal Replay Is Not a Simple Function of Experience," *Neuron* 65, no. 5 (March 2010): 695–705.

16. Zeb Kurth-Nelson et al., "Fast Sequences of Non-Spatial State Representations in Humans," *Neuron* 91, no. 1 (July 2016): 194–204; Yunzhe Liu et al., "Human Replay Spontaneously Reorganizes Experience," *Cell* 178, no. 3 (July 2019): 640–52.e14.

17. Daoyun Ji and Matthew A. Wilson, "Coordinated Memory Replay in the Visual Cortex and Hippocampus During Sleep," *Nature Neuroscience* 10, no. 1 (January 2007): 100–7.

18. Nicolas W. Schuck and Yael Niv, "Sequential Replay of Nonspatial Task States in the Human Hippocampus," *Science* 364, no. 6447 (June 2019).

19. Cameron Higgins et al., "Replay Bursts in Humans Coincide with Activation of the Default Mode and Parietal Alpha Networks," *Neuron* 109, no. 5 (March 2021): 882–93.e7.

20. Nikos K. Logothetis et al., "Hippocampal-Cortical Interaction During Periods of Subcortical Silence," *Nature* 491, no. 7425 (November 2012): 547–53.

4. NEURAL CIRCUITS OF THE HIPPOCAMPUS

1. Nikolaos Tzakis and Matthew R. Holahan, "Social Memory and the Role of the Hippocampal CA2 Region," *Frontiers in Behavioral Neuroscience* 13 (2019): 233; Andrew B. Lehr et al., "CA2 Beyond Social Memory: Evidence for a Fundamental Role in Hippocampal Information Processing," *Neuroscience & Biobehavioral Reviews* 126 (July 2021): 407–8.

2. There are excitatory (or principal) neurons as well as inhibitory neurons (or local interneurons) in the brain. There are also excitatory and inhibitory connections between

neurons. Here we consider only excitatory neurons and excitatory connections for the sake of simplicity. See David G. Amaral, Norio Ishizuka, and Brenda Claiborne, "Neurons, Numbers and the Hippocampal Network," *Progress in Brain Research* 83 (1990): 7–9.

3. David Marr, "Simple Memory: A Theory for Archicortex," *Philosophical Transactions of the Royal Society B: Biological Sciences* 262, no. 841 (July 1971): 23–81.

4. Donald O. Hebb, *The Organization of Behavior: A Psychological Theory* (New York: Wiley, 1949), 60–66.

5. Gyorgy Buzsaki, "Hippocampal Sharp Wave-Ripple: A Cognitive Biomarker for Episodic Memory and Planning," *Hippocampus* 25, no. 10 (October 2015): 1075–76.

6. CA1 is not completely without recurrent projections, but they are much weaker and directed differently compared to those of CA3. CA1 neurons project along the longitudinal axis (perpendicular to the cross-sectional plane) to connect with CA1 neurons in other cross-sections. Sunggu Yang et al., "Interlamellar CA1 Network in the Hippocampus," *Proceedings of the National Academy of Sciences of the United States* 111, no. 35 (September 2014): 12919–24.

5. VALUE-BASED DECISION-MAKING

1. Wolfram Schultz, Peter Dayan, and P. Read Montague, "A Neural Substrate of Prediction and Reward," *Science* 275, no. 5306 (March 1997): 1593–99.

2. Schultz, Dayan, and Montague, "A Neural Substrate of Prediction and Reward," 1593–99.

3. Wolfram Schultz et al., "Reward-Related Activity in the Monkey Striatum and Substantia Nigra," *Progress in Brain Research* 99 (1993): 227–35.

4. Daeyeol Lee, Hyojung Seo, and Min W. Jung, "Neural Basis of Reinforcement Learning and Decision Making," *Annual Review of Neuroscience* 35 (2012): 291–93; Camillo Padoa-Schioppa and Katherine E. Conen, "Orbitofrontal Cortex:A Neural Circuit for Economic Decisions," *Neuron* 96, no. 4 (November 2017): 739–42, 745–47.

5. Hyunjung Lee et al., "Hippocampal Neural Correlates for Values of Experienced Events," *Journal of Neuroscience* 32, no. 43 (October 2012): 15053–65; Sung-Hyun Lee et al., "Neural Signals Related to Outcome Evaluation Are Stronger in CA1 than CA3," *Frontiers in Neural Circuits* 11 (2017): 40; Eric B. Knudsen and Joni D. Wallis, "Hippocampal Neurons Construct a Map of an Abstract Value Space," *Cell* 184, no. 18 (September 2021): 4640–50 e10; Saori C. Tanaka et al., "Prediction of Immediate and Future Rewards Differentially Recruits Cortico-Basal Ganglia Loops," *Nature Neuroscience* 7, no. 8 (August 2004): 887–93; Katherine Duncan et al., "More Than the Sum of Its Parts: A Role for the Hippocampus in Configural Reinforcement Learning," *Neuron* 98, no. 3 (May 2018): 645–57.

6. Lee, Seo, and Jung, "Neural Basis of Reinforcement Learning and Decision Making," 291–93; Eun J. Shin et al., "Robust and Distributed Neural Representation of Action Values," *eLife* 10 (April 2021): e53045.

7. Lee, Seo, and Jung, "Neural Basis of Reinforcement Learning and Decision Making," 291–93.

8. The suprachiasmatic nucleus of the hypothalamus controls our physiology and behavior in accordance with the twenty-four-hour cycle. It is considered the master clock for the entire body.

9. Lee et al., "Hippocampal Neural Correlates for Values of Experienced Events," 15053–65.
10. Lee et al., "Neural Signals Related to Outcome Evaluation Are Stronger in CA1 than CA3," 40.
11. Robert J. McDonald and Norman M. White, "A Triple Dissociation of Memory Systems: Hippocampus, Amygdala, and Dorsal Striatum," *Behavioral Neuroscience* 107, no. 1 (February 1993): 15–18; Mark G. Packard and Barbara J. Knowlton, "Learning and Memory Functions of the Basal Ganglia," *Annual Review of Neuroscience* 25 (2002): 579–83.
12. Yeongseok Jeong et al., "Role of the Hippocampal CA1 Region in Incremental Value Learning," *Scientific Reports* 8, no. 1 (June 2018): 9870.
13. This is a technique commonly used in neuroscience to activate or silence specific types of cells in the brain. Synthetic proteins are expressed in target cells in a target brain area. The activation of the synthetic proteins by a compound (usually given to an animal by systemic injection) activates or turns off synthetic protein-expressing cells selectively.

6. REMEMBERING REWARDING FUTURES

1. Min W. Jung et al., "Remembering Rewarding Futures: A Simulation-Selection Model of the Hippocampus, " *Hippocampus* 28, no. 12 (December 2018): 913–30.
2. Sunggu Yang et al., "Interlamellar CA1 Network in the Hippocampus," *Proceedings of the National Academy of Sciences of the United States* 111, no. 35 (September 2014): 12919–24.
3. Gyorgy Buzsaki, "Hippocampal Sharp Wave-Ripple: A Cognitive Biomarker for Episodic Memory and Planning," *Hippocampus* 25, no. 10 (October 2015): 1075–76.
4. Federico Stella et al., "Hippocampal Reactivation of Random Trajectories Resembling Brownian Diffusion," *Neuron* 102, no. 2 (April 2019): 450–61.
5. Matthijs van der Meer, Zeb Kurth-Nelson, and A. David Redish, "Information Processing in Decision-Making Systems," *Neuroscientist* 18, no. 4 (August 2012): 352–54; Giovanni Pezzulo et al., "Internally Generated Sequences in Learning and Executing Goal-Directed Behavior," *Trends in Cognitive Sciences* 18, no. 12 (December 2014): 652–54; Andrew M. Wikenheiser and Geoffrey Schoenbaum, "Over the River, through the Woods: Cognitive Maps in the Hippocampus and Orbitofrontal Cortex," *Nature Reviews Neuroscience* 17, no. 8 (August 2016): 521.
6. Testing replays of spatial trajectories requires simultaneous recording of a sufficiently large number of place cells (on the order of ten). However, monitoring single place cell activity is sufficient to test reactivation during sharp-wave ripples (i.e., the degree to which a particular place cell is active together with sharp-wave ripples).
7. Annabelle C. Singer and Loren M. Frank, "Rewarded Outcomes Enhance Reactivation of Experience in the Hippocampus," *Neuron* 64, no. 6 (December 2009): 910–21; David Dupret et al., "The Reorganization and Reactivation of Hippocampal Maps Predict Spatial Memory Performance," *Nature Neuroscience* 13, no. 8 (August 2010): 995–1002.
8. Brad E. Pfeiffer, and David J. Foster, "Hippocampal Place-Cell Sequences Depict Future Paths to Remembered Goals." *Nature* 497, no. 7447 (May 2013): 74–79; H. Freyja Olafsdottir et al., "Hippocampal Place Cells Construct Reward Related Sequences through

Unexplored Space,'" *eLife* 4 (June 2015): e06063; R. Ellen Ambrose, Brad E. Pfeiffer, and David J. Foster, "Reverse Replay of Hippocampal Place Cells Is Uniquely Modulated by Changing Reward," *Neuron* 91, no. 5 (September 2016): 1124–36; Baburam Bhattarai, Jong W. Lee, and Min W. Jung, "Distinct Effects of Reward and Navigation History on Hippocampal Forward and Reverse Replays,'" *Proceedings of the National Academy of Sciences of the United States* 117, no. 1 (January 2020): 689–97.

9. Lisa Bulganin and Bianca C. Wittmann, "Reward and Novelty Enhance Imagination of Future Events in a Motivational-Episodic Network," *PLoS One* 10, no. 11 (2015): e0143477; Matthias J. Gruber et al., "Post-learning Hippocampal Dynamics Promote Preferential Retention of Rewarding Events," *Neuron* 89, no. 5 (March 2016): 1110–20.

10. Jung et al., "Remembering Rewarding Futures," 913–30.

11. Bruce L. McNaughton, "Neuronal Mechanisms for Spatial Computation and Information Storage," in *Neural Connections, Mental Computations*, ed. Lynn Nadel et al. (Cambridge, MA: MIT Press, 1989), 305; Edmund T. Rolls, "Functions of Neuronal Networks in the Hippocampus and Cerebral Cortex in Memory," in *Models of Brain Function*, ed. Rodney M. J. Cotterill (Cambridge: Cambridge University Press, 1989), 18–21; James J. Knierim, and Joshua Neunuebel, "Tracking the Flow of Hippocampal Computation: Pattern Separation, Pattern Completion, and Attractor Dynamics," *Neurobiology of Learning and Memory* 129 (March 2016): 39–46.

12. Jong W. Lee and Min W. Jung, "Separation or Binding? Role of the Dentate Gyrus in Hippocampal Mnemonic Processing," *Neuroscience & Biobehavioral Reviews* 75 (April 2017): 183–91.

13. Richard S. Sutton, "Dyna, an Integrated Architecture for Learning, Planning, and Reacting," *ACM Sigart Bulletin* 2, no. 4 (1991): 160–63377; Richard S. Sutton and Andrew G. Barto, *Reinforcement Learning: An Introduction* (Cambridge, MA: MIT Press, 1998), 230–35.

7. THE EVOLUTION OF IMAGINATION

1. Joseph R. Manns and Howard Eichenbaum, "Evolution of Declarative Memory," *Hippocampus* 16, no. 9 (2006): 796–98.

2. Verner P. Bingman, Cosme Salas, and Fernando Rodríguez, "Evolution of the Hippocampus," in *Encyclopaedia of Neuroscience*, ed. Marc D. Binder, Nobutaka Hirokawa and Uwe Windhorst (Berlin: Springer, 2009), 1356–60; Christina Herold, Vincent J. Coppola, and Verner P. Bingman, "The Maturation of Research into the Avian Hippocampal Formation: Recent Discoveries from One of the Nature's Foremost Navigators," *Hippocampus* 25, no. 11 (November 2015): 1194–200; Georg F. Striedter, "Evolution of the Hippocampus in Reptiles and Birds," *Journal of Comparative Neurology* 524, no. 3 (February 15 2016): 497–507.

3. Sara J. Shettleworth, "Spatial Memory in Food-Storing Birds," *Philosophical Transactions of the Royal Society B: Biological Sciences* 329, no. 1253 (1990): 143–51.

4. Jennifer J. Siegel, Douglas Nitz, and Verner P. Bingman, "Lateralized Functional Components of Spatial Cognition in the Avian Hippocampal Formation: Evidence from Single-Unit Recordings in Freely Moving Homing Pigeons," *Hippocampus* 16, no. 2 (2006): 125–40; Jennifer J. Siegel, Douglas Nitz, and Verner P. Bingman, "Spatial-Specificity of Single-Units in the Hippocampal Formation of Freely Moving Homing Pigeons," *Hippocampus* 15, no. 1 (2005): 26–40.

5. Randy L. Buckner, Jessica R. Andrews-Hanna, and Daniel L. Schacter, "The Brain's Default Network: Anatomy, Function, and Relevance to Disease," *Annals of the New York Academy of Sciences* 1124 (March 2008): 20–24, 30.

6. Nina Patzke et al., "In Contrast to Many Other Mammals, Cetaceans Have Relatively Small Hippocampi That Appear to Lack Adult Neurogenesis," *Brain Structure and Function* 220, no. 1 (January 2015): 361–83.

7. Maya Geva-Sagiv et al., "Spatial Cognition in Bats and Rats: From Sensory Acquisition to Multiscale Maps and Navigation," *Nature Reviews Neuroscience* 16, no. 2 (February 2015): 96, 101–2.

8. T. A. Stevens and J. R. Krebs, "Retrieval of Stored Seeds by Marsh Tits *Parus Palustris* in the Field," *Ibis* 128, no. 4 (1986): 513–25.

9. Hannah L. Payne, Galen F. Lynch, and Dmitriy Aronov, "Neural Representations of Space in the Hippocampus of a Food-Caching Bird," *Science* 373, no. 6552 (July 2021): 343–48,

10. Elhanan Ben-Yishay et al., "Directional Tuning in the Hippocampal Formation of Birds," *Current Biology* 31, no. 12 (June 2021): 2592–602.e4.

8. ABSTRACT THINKING AND NEOCORTEX

1. Howard Eichenbaum et al., "The Hippocampus, Memory, and Place Cells: Is It Spatial Memory or a Memory Space?" *Neuron* 23, no. 2 (June 1999): 213–15.

2. Charles R. Gallistel, *The Organization of Learning* (Cambridge, MA: MIT Press, 1990), 338–40; Sara J. Shettleworth, *Cognition, Evolution, and Behavior*, 2nd ed. (Oxford: Oxford University Press, 2010), 190–208, 421–55; Christopher Krupenye and Josep Call, "Theory of Mind in Animals: Current and Future Directions," *Wiley Interdisciplinary Reviews: Cognitive Science* 10, no. 6 (November 2019): e1503; Caio A. Lage, De Wet Wolmarans, and Daniel C. Mograbi, "An Evolutionary View of Self-Awareness," *Behavioural Processes* 194 (January 2022): 104543.

3. Carol A. Seger and Earl K. Miller, "Category Learning in the Brain," *Annual Review of Neuroscience* 33 (2010): 205–9; Raymond P. Kesner and John C. Churchwell, "An Analysis of Rat Prefrontal Cortex in Mediating Executive Function," *Neurobiology of Learning and Memory* 96, no. 3 (October 2011): 422–23; Sébastien Tremblay, K. M. Sharika, and Michael L. Platt, "Social Decision-Making and the Brain: A Comparative Perspective," *Trends in Cognitive Sciences* 21, no. 4 (April 2017): 269–70; Farshad A. Mansouri, David J. Freedman, and Mark J. Buckley, "Emergence of Abstract Rules in the Primate Brain," *Nature Reviews Neuroscience* 21, no. 11 (November 2020): 597–602; Prabaha Gangopadhyay et al., "Prefrontal-Amygdala Circuits in Social Decision-Making," *Nature Neuroscience* 24, no. 1 (January 2023): 5–13.

4. Immanuel Kant, *Critique of Pure Reason*, ed. Paul Guyer and Allen W. Wood (Cambridge: Cambridge University Press, 1999), 110, 127–29, 136–38.

5. Kant, *Critique of Pure Reason*, 157–59, 178–80.

6. Torkel Hafting et al., "Microstructure of a Spatial Map in the Entorhinal Cortex," *Nature* 436, no. 7052 (August 2005): 801–6.

7. Eric L. Hargreaves et al., "Major Dissociation between Medial and Lateral Entorhinal Input to Dorsal Hippocampus," *Science* 308, no. 5729 (June 2005): 1792–94.

8. Alexandra O. Constantinescu, Jill X. O'Reilly, and Timothy E. J. Behrens, "Organizing Conceptual Knowledge in Humans with a Gridlike Code," *Science* 352, no. 6292 (June 2016): 1464–68; Seongmin A. Park, Douglas S. Miller, and Erie D. Boorman,

"Inferences on a Multidimensional Social Hierarchy Use a Grid-Like Code," *Nature Neuroscience* 24, no. 9 (September 2021): 1292–301.

9. Timothy E. J. Behrens et al., "What Is a Cognitive Map? Organizing Knowledge for Flexible Behavior," *Neuron* 100, no. 2 (October 2018): 502–4.

10. Kant, *Critique of Pure Reason*, 127–29, 136–38.

11. Gilbert Ryle, *The Concept of Mind* (New York: Barnes and Noble, 1949), 16.

12. "Gilbert Ryle," Wikipedia, updated October 28, 2021, https://en.wikipedia.org/wiki/Gilbert_Ryle.

13. Note that the cerebellum is also involved in a wide range of cognitive tasks. Richard B. Ivry and Juliana V. Baldo, "Is the Cerebellum Involved in Learning and Cognition?" *Current Opinion in Neurobiology* 2, no. 2 (1992): 214; Maedbh King et al., "Functional Boundaries in the Human Cerebellum Revealed by a Multi-Domain Task Battery," *Nature Neuroscience* 22, no. 8 (2019): 1371–78.

14. Suzana Herculano-Houzel et al., "The Elephant Brain in Numbers," *Frontiers in Neuroanatomy* 8 (2014): 46.

15. David C. Van Essen, Chad J. Donahue, and Matthew F. Glasser, "Development and Evolution of Cerebral and Cerebellar Cortex," *Brain, Behavior and Evolution* 91, no. 3 (2018): 159; Suzana Herculano-Houzel, "The Human Brain in Numbers: A Linearly Scaled-Up Primate Brain," *Frontiers in Human Neuroscience* 3 (2009): 31.

16. Herculano-Houzel, "The Human Brain in Numbers," 31.

17. Cornelia McCormick et al., "Mind-Wandering in People with Hippocampal Damage," *Journal of Neuroscience* 38, no. 11 (March 2018): 2745–54.

18. Jon H. Kaas, "The Origin and Evolution of Neocortex: From Early Mammals to Modern Humans," *Progress in Brain Research* 250 (2019): 72–73.

19. Ferris Jabr, "How Humans Evolved Supersize Brains," *Quanta Magazine*, November 10, 2015, https://www.quantamagazine.org/how-humans-evolved-supersize-brains-20151110/.

20. Kaas, "The Origin and Evolution of Neocortex," 64.

21. Bradley L. Schlaggar and Dennis D. O'Leary, "Potential of Visual Cortex to Develop an Array of Functional Units Unique to Somatosensory Cortex," *Science* 252, no. 5012 (June 1991): 1556–60.

22. Leslie C. Aiello and Peter Wheeler, "The Expensive-Tissue Hypothesis: The Brain and the Digestive System in Human and Primate Evolution," *Current Anthropology* 36, no. 2 (1995): 201–8.

23. Rachel N. Carmody and Richard W. Wrangham, "The Energetic Significance of Cooking," *Journal of Human Evolution* 57, no. 4 (October 2009): 380–86; Karina Fonseca-Azevedo and Suzana Herculano-Houzel, "Metabolic Constraint Imposes Tradeoff between Body Size and Number of Brain Neurons in Human Evolution," *Proceedings of the National Academy of Sciences of the United States* 109, no. 45 (November 2012): 18571–6.

24. Alianda M. Cornelio et al., "Human Brain Expansion During Evolution Is Independent of Fire Control and Cooking," *Frontiers in Neuroscience* 10 (2016): 167.

25. Yuval Noah Harari, *Sapiens: A Brief History of Humankind*, trans. Hyun-Wook Cho (Paju, South Korea: Gimm-Young, 2015), 50–60.

9. PREFRONTAL CORTEX

1. Chad J. Donahue et al., "Quantitative Assessment of Prefrontal Cortex in Humans Relative to Nonhuman Primates," *Proceedings of the National Academy of Sciences of the United States* 115, no. 22 (May 2018): E5183–E92.

2. John M. Harlow, "Recovery from the Passage of an Iron Bar through the Head," *Publication of the Massachusetts Medical Society* 2, no. 3 (1868): 329–47

3. Tim Shallice and Lisa Cipolotti, "The Prefrontal Cortex and Neurological Impairments of Active Thought," *Annual Review of Psychology* 69 (January 2018): 169.

4. Patrick Murphy et al., "Impairments in Proverb Interpretation Following Focal Frontal Lobe Lesions," *Neuropsychologia* 51, no. 11 (September 2013): 2075–86.

5. Andreas Nieder, "Supramodal Numerosity Selectivity of Neurons in Primate Prefrontal and Posterior Parietal Cortices," *Proceedings of the National Academy of Sciences of the United States* 109, no. 29 (July 2012): 11860–5.

6. Andreas Nieder, "The Neuronal Code for Number," *Nature Reviews Neuroscience* 17, no. 6 (2016): 367–69.

7. Numerous studies, such as the following, have found neural activity related to abstract concepts in the rodent prefrontal cortex: Erin L. Rich and Matthew Shapiro, "Rat Prefrontal Cortical Neurons Selectively Code Strategy Switches," *Journal of Neuroscience* 29, no. 22 (June 2009): 7208–19; Gregory B. Bissonette and Matthew R. Roesch, "Neural Correlates of Rules and Conflict in Medial Prefrontal Cortex During Decision and Feedback Epochs," *Frontiers in Behavioral Neuroscience* 9 (2015): 266; James M. Hyman et al., "Contextual Encoding by Ensembles of Medial Prefrontal Cortex Neurons," *Proceedings of the National Academy of Sciences of the United States* 109, no. 13 (March 2012): 5086–91; Sandra Reinert et al., "Mouse Prefrontal Cortex Represents Learned Rules for Categorization," *Nature* 593, no. 7859 (May 2021): 411–17.

8. David S. Tait, E. Alexander Chase, and Verity J. Brown, "Attentional Set-Shifting in Rodents: A Review of Behavioural Methods and Pharmacological Results," *Current Pharmaceutical Design* 20, no. 31 (2014): 5046–59.

9. Iroise Dumontheil, "Development of Abstract Thinking During Childhood and Adolescence: The Role of Rostrolateral Prefrontal Cortex," *Developmental Cognitive Neuroscience* 10 (October 2014): 64–69.

10. Dumontheil, "Development of Abstract Thinking During Childhood and Adolescence," 59; David Badre, "Cognitive Control, Hierarchy, and the Rostro-Caudal Organization of the Frontal Lobes," *Trends in Cognitive Science* 12, no. 5 (May 2008): 194–95.

10. THE HUMAN REVOLUTION AND ASSOCIATED BRAIN CHANGES

1. Ofer Bar-Yosef, "The Upper Paleolithic Revolution," *Annual Review of Anthropology* 31, no. 1 (2002): 364–69; Pamela R. Willoughby, "Modern Human Behavior," in *Oxford Research Encyclopedia of Anthropology*, published online May 29, 2020, https://doi.org/10.1093/acrefore/9780190854584.013.46.

2. Gillian M. Morriss-Kay, "The Evolution of Human Artistic Creativity," *Journal of Anatomy* 216, no. 2 (February 2010): 166.

3. Morriss-Kay, "The Evolution of Human Artistic Creativity," 171; Dahlia W. Zaidel, "Art and Brain: The Relationship of Biology and Evolution to Art," *Progress in Brain Research* 204 (2013): 218, 223.

4. Julien Riel-Salvatore and Claudine Gravel-Miguel, "Upper Palaeolithic Mortuary Practices in Eurasia: A Critical Look at the Burial Record," in *The Oxford Handbook of the Archaeology of Death and Burial*, ed. Liv N. Stutz and Sarah Tarlow (Oxford: Oxford University Press, 2013), 303, 335–36.

5. Judith Thurman, "First Impressions: What Does the World's Oldest Art Say about Us?" *New Yorker*, June 16, 2008.
6. Takanori Kochiyama et al., "Reconstructing the Neanderthal Brain Using Computational Anatomy," *Scientific Reports* 8, no. 1 (April 2018): 6296; Axel Timmermann, "Quantifying the Potential Causes of Neanderthal Extinction: Abrupt Climate Change Versus Competition and Interbreeding," *Quaternary Science Reviews* 238 (2020): 106331.
7. R. Gabriel Joseph, *Frontal Lobes: Neuroscience, Neuropsychology, Neuropsychiatry* (Cambridge, MA: Cosmology, 2011), 190–99.
8. Heléne Valladas et al., "Thermoluminescence Dating of Mousterian Troto-Cro-Magnon Remains from Israel and the Origin of Modern Man," *Nature* 331, no. 6157 (1988): 614–16.
9. Ian McDougall, Francis H. Brown, and John G. Fleagle, "Stratigraphic Placement and Age of Modern Humans from Kibish, Ethiopia," *Nature* 433, no. 7027 (2005): 733–36 Jean-Jacques Hublin et al., "New Fossils from Jebel Irhoud, Morocco and the Pan-African Origin of *Homo Sapiens*," *Nature* 546, no. 7657 (June 2017): 289–92.
10. Rebecca L. Cann, Mark Stoneking, and Allan C. Wilson, "Mitochondrial DNA and Human Evolution," *Nature* 325, no. 6099 (1987): 31–36; Douglas C. Wallace, Michael D. Brown, and Marie T. Lott, "Mitochondrial DNA Variation in Human Evolution and Disease," *Gene* 238, no. 1 (September 1999): 211–30; Renée Hetherington and Robert G. B. Reid, *The Climate Connection: Climate Change and Modern Human Evolution* (Cambridge: Cambridge University Press, 2010), 31–34.
11. Genetic evidence indicates that *Homo sapiens* interbred with other early humans including Neanderthals rather than completely replacing them. Philipp Gunz et al., "Neandertal Introgression Sheds Light on Modern Human Endocranial Globularity," *Current Biology* 29, no. 1 (January 2019): 120–27.
12. Richard G. Klein, "Out of Africa and the Evolution of Human Behavior," *Evolutionary Anthropology: Issues, News, and Reviews* 17, no. 6 (2008): 271.
13. Adam Powell, Stephen Shennan, and Mark G. Thomas, "Late Pleistocene Demography and the Appearance of Modern Human Behavior," *Science* 324, no. 5932 (June 2009): 1298–301.
14. Stephen Davies, "Behavioral Modernity in Retrospect," *Topoi* 40, no. 1 (2021): 228–29.
15. Amélie Beaudet, Andrew Du, and Bernard Wood, "Evolution of the Modern Human Brain," *Progress in Brain Research* 250 (2019): 229–41.
16. Simon Neubauer, Jean-Jacques Hublin, and Philipp Gunz, "The Evolution of Modern Human Brain Shape," *Science Advances* 4, no. 1 (2018): eaao5961.
17. Neubauer, Hublin, and Gunz, "The Evolution of Modern Human Brain Shape," eaao5961.
18. Beaudet, Du, and Wood, "Evolution of the Modern Human Brain," 228–29; Emiliano Bruner et al., "The Brain and the Braincase: A Spatial Analysis on the Midsagittal Profile in Adult Humans," *Journal of Anatomy* 227, no. 3 (September 2015): 275.
19. Emiliano Bruner et al., "Evidence for Expansion of the Precuneus in Human Evolution," *Brain Structure and Function* 222, no. 2 (March 2017): 1053–60.
20. Emiliano Bruner and Sofia Pereira-Pedro, "A Metric Survey on the Sagittal and Coronal Morphology of the Precuneus in Adult Humans," *Brain Structure and Function* 225, no. 9 (December 2020): 2747–55; Emiliano Bruner et al., "Cortical Surface Area and Cortical Thickness in the Precuneus of Adult Humans," *Neuroscience* 286 (Feb 12 2015): 345–52.

21. Emiliano Bruner, "Human Paleoneurology and the Evolution of the Parietal Cortex," *Brain, Behavior and Evolution* 91, no. 3 (2018): 143–44.
22. Andrea E. Cavanna and Michael R. Trimble, "The Precuneus: A Review of Its Functional Anatomy and Behavioural Correlates," *Brain* 129, part 3 (March 2006): 568–78.
23. The gray matter is made up of neuronal cell bodies, dendrites, and axons, and it is responsible for processing information. The white matter is made up of nerve fibers that are coated in a fatty substance called myelin, which allows for faster electrical transmission of signals and connects different areas of the brain to each other.
24. Rebecca Chamberlain et al., "Drawing on the Right Side of the Brain: A Voxel-Based Morphometry Analysis of Observational Drawing," *Neuroimage* 96 (August 2014): 167–73.
25. Shoji Tanaka and Eiji Kirino, "Reorganization of the Thalamocortical Network in Musicians," *Brain Research* 1664 (June 2017): 48–54.
26. Qun-Lin Chen et al., "Individual Differences in Verbal Creative Thinking Are Reflected in the Precuneus," *Neuropsychologia* 75 (August 2015): 441–49.
27. Emanuel Jauk et al., "Gray Matter Correlates of Creative Potential: A Latent Variable Voxel-Based Morphometry Study," *Neuroimage* 111 (May 2015): 312–20.
28. Debra A. Gusnard and Marcus E. Raichle, "Searching for a Baseline: Functional Imaging and the Resting Human Brain," *Nature Reviews Neuroscience* 2, no. 10 (October 2001): 690.
29. Amanda V. Utevsky, David V. Smith, and Scott A. Huettel, "Precuneus Is a Functional Core of the Default-Mode Network," *Journal of Neuroscience* 34, no. 3 (January 2014): 932–40.
30. Reece P. Roberts and Donna Rose Addis, "A Common Mode of Processing Governing Divergent Thinking and Future Imagination," in *The Cambridge Handbook of the Neuroscience of Creativity*, ed. Rex E. Jung and Oshin Vartanian (Cambridge: Cambridge University Press, 2018), 213–15; Roger E. Beaty, "The Creative Brain," *Cerebrum* (January 2020): cer-02-20; Oshin Vartanian, "Neuroscience of Creativity," in *The Cambridge Handbook of Creativity*, ed. James C. Kaufman and Robert J. Sternberg (Cambridge: Cambridge University Press, 2019), 156–59.
31. Jared Diamond, "The Great Leap Forward," in *Technology and Society: Issue for the 21st Century and Beyond*, 3rd ed., ed. Linda Hjorth, Barbara Eichler, Ahmed Khan and John Morello (Upper Saddle River, NJ: Prentice Hall, 2008), 15.

11. DEEP NEURAL NETWORK

1. Yann LeCun, Yoshua Bengio, and Geoffrey Hinton, "Deep Learning," *Nature* 521, no. 7553 (May 2015): 436; Alexander S. Lundervold and Arvid Lundervold, "An Overview of Deep Learning in Medical Imaging Focusing on MRI," *Zeitschrift für Medizinische Physik* 29, no. 2 (May 2019): 102–3; Roger Parloff, "Why Deep Learning is Suddenly Changing Your Life," *Fortune*, September 29, 2016, https://fortune.com/longform/ai-artificial-intelligence-deep-machine-learning/.
2. Hannes Schulz and Sven Behnke, "Deep Learning: Layer-Wise Learning of Feature Hierarchies," *Künstliche Intelligenz* 26, no. 4 (2012): 357–63.
3. Quoc V. Le et al., "Building High-Level Features Using Large Scale Unsupervised Learning" (paper presented at the Proceedings of the 29th International Conference on Machine Learning, 2012), https://doi.org/10.48550/arXiv.1112.6209.

4. Gwangsu Kim et al., "Visual Number Sense in Untrained Deep Neural Networks," *Science Advances* 7, no. 1 (2021): eabd6127.

5. Seungdae Baek et al., "Face Detection in Untrained Deep Neural Networks," *Nature Communications* 12, no. 1 (Dec 16 2021): 7328.

6. Jon H. Kaas, "The Origin and Evolution of Neocortex: From Early Mammals to Modern Humans," *Progress in Brain Research* 250 (2019): 64.

7. James J. DiCarlo, Davide Zoccolan, and Nicole C. Rust, "How Does the Brain Solve Visual Object Recognition?" *Neuron* 73, no. 3 (February 2012): 419–20; Jon H. Kaas and Mary K. L. Baldwin, "The Evolution of the Pulvinar Complex in Primates and Its Role in the Dorsal and Ventral Streams of Cortical Processing," *Vision* 4, no. 1 (December 2019): 3; Sang-Han Choi et al., "Proposal for Human Visual Pathway in the Extrastriate Cortex by Fiber Tracking Method Using Diffusion-Weighted MRI," *Neuroimage* 220 (October 2020): 117145; Joshua H. Siegle et al., "Survey of Spiking in the Mouse Visual System Reveals Functional Hierarchy," *Nature* 592, no. 7852 (April 2021): 86–92.

8. Hernan G. Rey et al., "Single Neuron Coding of Identity in the Human Hippocampal Formation," *Current Biology* 30, no. 6 (March 2020): 1152–59.

9. Rodrigo Q. Quiroga, "Concept Cells: The Building Blocks of Declarative Memory Functions," *Nature Reviews Neuroscience* 13, no. 8 (July 2012): 587–91; Rodrigo Q. Quiroga et al., "Invariant Visual Representation by Single Neurons in the Human Brain," *Nature* 435, no. 7045 (June 2005): 1102–7.

10. Julia Sliwa et al., "Independent Neuronal Representation of Facial and Vocal Identity in the Monkey Hippocampus and Inferotemporal Cortex," *Cerebral Cortex* 26, no. 3 (March 2016): 950–66.

11. Thérèse M. Jay, Jacques Glowinski, and Anne-Marie Thierry, "Selectivity of the Hippocampal Projection to the Prelimbic Area of the Prefrontal Cortex in the Rat," *Brain Research* 505, no. 2 (December 1989): 337–40; Priyamvada Rajasethupathy et al., "Projections from Neocortex Mediate Top-Down Control of Memory Retrieval," *Nature* 526, no. 7575 (October 2015): 653–59; Ruchi Malik et al., "Top-Down Control of Hippocampal Signal-to-Noise by Prefrontal Long-Range Inhibition," *Cell* 185, no. 9 (April 2022): 1602–17.

12. Michel A. Hofman, "Evolution of the Human Brain: When Bigger Is Better," *Frontiers in Neuroanatomy* 8 (2014): 15.

13. Lei Xing et al., "Expression of Human-Specific *ARHGAP11B* in Mice Leads to Neocortex Expansion and Increased Memory Flexibility," *EMBO Journal* 40, no. 13 (July 2021): e107093.

14. Michael Heide et al., "Human-Specific *ARHGAP11B* Increases Size and Folding of Primate Neocortex in the Fetal Marmoset," *Science* 369, no. 6503 (July 2020): 546–50.

12. SHARING IDEAS AND KNOWLEDGE THROUGH LANGUAGE

1. Mark Pagel, "Q&A: What Is Human Language, When Did It Evolve and Why Should We Care?" *BMC Biology* 15, no. 1 (July 2017): 64.

2. Pagel, "Q&A," 64.

3. Marc D. Hauser et al., "The Mystery of Language Evolution," *Frontiers in Psychology* (2014): 401; Willem Zuidema and Arie Verhagen, "What Are the Unique Design

Features of Language? Formal Tools for Comparative Claims," *Adaptive Behavior* 18, no. 1 (2010): 49–51; Sławomir Wacewicz and Przemysław Żywiczyński, "Language Evolution: Why Hockett's Design Features Are a Non-Starter," *Biosemiotics* 8, no. 1 (2015): 30–35; Michael D. Beecher, "Why Are No Animal Communication Systems Simple Languages?" *Frontiers in Psychology* (2021): 701; R. Haven Wiley, "Design Features of Language," in *Encyclopedia of Evolutionary Psychological Science* (Berlin: Springer, 2021), 1922–28; Charles F. Hockett, "Animal "Languages" and Human Language," *Human Biology* 31, no. 1 (1959): 32–36; Charles F. Hockett and Charles D. Hockett, "The Origin of Speech," *Scientific American* 203, no. 3 (1960): 90–92.

4. Hiroyuki Ai et al., "Neuroethology of the Waggle Dance: How Followers Interact with the Waggle Dancer and Detect Spatial Information," *Insects* 10, no. 10 (October 2019): 336; Randolf Menzel, "The Waggle Dance as an Intended Flight: A Cognitive Perspective," *Insects* 10, no. 12 (November 2019): 424

5. Patricia Dennis, Stephen M. Shuster, and C. N. Slobodchikoff, "Dialects in the Alarm Calls of Black-Tailed Prairie Dogs (*Cynomys ludovicianus*): A Case of Cultural Diffusion?" *Behavioral Processes* 181 (December 2020): 104243; Con N. Slobodchikoff, Andrea Paseka, and Jennifer L. Verdolin, "Prairie Dog Alarm Calls Encode Labels about Predator Colors," *Animal Cognition* 12, no. 3 (May 2009): 435–39.

6. Francine Patterson and Wendy Gordon, "The Case for the Personhood of Gorillas," in *The Great Ape Project*, ed. Paola Cavalieri and Peter Singer (New York: St. Martin's, 1993), 59.

7. Francine Patterson and Wendy Gordon, "Twenty-Seven Years of Project Koko and Michael," in *All Apes Great and Small* (Berlin: Springer, 2002), 168.

8. Patterson and Gordon, "The Case for the Personhood of Gorillas," 67.

9. Laasya Samhita, and Hans J. Gross, "The 'Clever Hans Phenomenon' Revisited," *Communicative & Integrative Biology* 6, no. 6 (November 2013): e27122.

10. Patterson and Gordon, "The Case for the Personhood of Gorillas," 62.

11. Geoffrey K. Pullum, "Koko Is Dead, but the Myth of Her Linguistic Skills Lives On," The Chronicle of Higher Education, June 27, 2018, https://www.chronicle.com/blogs /linguafranca/koko-is-dead-but-the-myth-of-her-linguistic-skills-lives-on.

12. Ramirez et al., "Creating a False Memory in the Hippocampus," *Science* 341, no. 6144 (July 2013): 387–91.

13. Maria Konnikova, "The Man Who Couldn't Speak and How He Revolutionized Psychology," *Scientific American*, February 3, 2018, https://blogs.scientificamerican .com/literally-psyched/the-man-who-couldnt-speakand-how-he-revolutionized -psychology/; Nina F. Dronkers et al., "Paul Broca's Historic Cases: High Resolution MR Imaging of the Brains of Leborgne and Lelong," *Brain* 130, no. 5 (2007): 1432–41.

14. Mario Lanczik and G. Keil, "Carl Wernicke's Localization Theory and Its Significance for the Development of Scientific Psychiatry," *History of Psychiatry* 2, no. 6 (1991): 174–75; William F. Bynum, "Wernicke, Carl," Encyclopedia.com, updated May 8, 2018, https:// www.encyclopedia.com/people/medicine/psychology-and-psychiatry-biographies /carl-wernicke.

15. David Poeppel et al., "Towards a New Neurobiology of Language," *Journal of Neuroscience* 32, no. 41 (October 2012): 14125–26; Peter Hagoort, "MUC (Memory, Unification, Control) and Beyond," *Frontiers in Psychology* 4 (2013): 416.

16. Poeppel et al., "Towards a New Neurobiology of Language," 14126.

17. Poeppel et al., "Towards a New Neurobiology of Language," 14130.

18. Oscar Wilde, "The Critic as Artist," 1891, accessed November 11, 2022, https://celt.ucc .ie/published/E800003-007/text001.html, emphasis added.

19. Ludwig Wittgenstein, *Tractatus Logico-Philosophicus*, trans. C K. Ogden (London: Routledge, 1981), 149.

20. Bertrand Russell, *Human Knowledge: Its Scope and Limits* (London: Routledge, 2009), 74.

21. Lila Gleitman and Anna Papafragou, "Language and Thought," in *Cambridge Handbook of Thinking and Reasoning*, ed. Keith Holyoak and Robert G. Morrison (Cambridge: Cambridge University Press, 2005), 634.

22. Paul Bloom, "Language and Thought: Does Grammar Makes Us Smart?" *Current Biology* 10, no. 14 (2000): R516-R17; Michael Siegal and Rosemary Varley, "Aphasia, Language, and Theory of Mind," *Social Neuroscience* 1, no. 3–4 (2006): 169–70; Rosemary Varley and Michael Siegal, "Evidence for Cognition without Grammar from Causal Reasoning and 'Theory of Mind' in an Agrammatic Aphasic Patient," *Current Biology* 10, no. 12 (2000): 723–26; Rosemary A. Varley et al., "Agrammatic but Numerate," *Proceedings of the National Academy of Sciences of the United States* 102, no. 9 (2005): 3519–24.

23. Michael Rescorla, "The Language of Thought Hypothesis," *The Stanford Encyclopedia of Philosophy* (Summer 2019 ed.), accessed November 11, 2022, https://plato.stanford .edu/archives/sum2019/entries/language-thought/.

24. Paul Bloom and Frank. C Keil, "Thinking through Language," *Mind & Language* 16, no. 4 (2001): 364.

25. Basel A.-S. Hussein, "The Sapir-Whorf Hypothesis Today," *Theory and Practice in Language Studies* 2, no. 3 (March 2012): 642–45.

26. Guy Dove, "More Than a Scaffold: Language Is a Neuroenhancement," *Cognitive Neuropsychology* 37, no. 5–6 (2020): 288–89, .

27. Anna M. Borghi et al., "Words as Social Tools: Language, Sociality and Inner Grounding in Abstract Concepts," *Physics of Life Reviews* 29 (2019): 142.

28. Gary Lupyan and Bodo Winter, "Language Is More Abstract Than You Think, or, Why Aren't Languages More Iconic?" *Philosophical Transactions of the Royal Society B: Biological Sciences* 373, no. 1752 (2018): 20170137.

29. Lupyan and Winter, "Language Is More Abstract Than You Think?" 20170137.

30. Leonid Perlovsky, "Language and Cognition—Joint Acquisition, Dual Hierarchy, and Emotional Prosody," *Frontiers in Behavioral Neuroscience* 7 (2013): 123.

31. Bloom, "Language and Thought," R516-R17; Siegal and Varley, "Aphasia, Language, and Theory of Mind," 169–70; Varley and Siegal, "Evidence for Cognition Without Grammar," 723–26; Varley et al., "Agrammatic but Numerate," 3519–24; Peter Langland-Hassan et al., "Assessing Abstract Thought and Its Relation to Language with a New Nonverbal Paradigm: Evidence from Aphasia," *Cognition* 211 (2021): 104622; Peter Langland-Hassan et al., "Metacognitive Deficits in Categorization Tasks in a Population with Impaired Inner Speech," *Acta Psychologica* 181 (2017): 62–74.

32. Dove, "More Than a Scaffold," 289–90.

33. Neil Glickman, Charlene Crump, and Steve Hamerdinger, "Language Deprivation Is a Game Changer for the Clinical Specialty of Deaf Mental Health," *JADARA* 54, no. 1 (2020): 62–63; Matthew L. Hall et al., "Auditory Deprivation Does Not Impair Executive Function, but Language Deprivation Might: Evidence from a Parent-Report Measure in Deaf Native Signing Children," *The Journal of Deaf Studies and*

Deaf Education 22, no. 1 (2017): 9–21; Wyatte C. Hall, Leonard L. Levin, and Melissa L Anderson, "Language Deprivation Syndrome: A Possible Neurodevelopmental Disorder with Sociocultural Origins," *Social Psychiatry and Psychiatric Epidemiology* 52, no. 6 (2017): 767; Sanjay Gulati, "Language Deprivation Syndrome," in *Language Deprivation and Deaf Mental Health*, ed. Neil S. Glickman and Wyatte C. Hall (London: Routledge, 2018), 32–34.

34. Qi Cheng et al., "Effects of Early Language Deprivation on Brain Connectivity: Language Pathways in Deaf Native and Late First-Language Learners of American Sign Language," *Frontiers in Human Neuroscience* 13 (2019): 320; Rachel I. Mayberry et al., "Age of Acquisition Effects on the Functional Organization of Language in the Adult Brain," *Brain and Language* 119, no. 1 (2011): 16–29.

35. Gulati, "Language Deprivation Syndrome," 34.

36. Siegal and Varley, "Aphasia, Language, and Theory of Mind," 172.

37. Louis-Jean Boë et al., "Which Way to the Dawn of Speech? Reanalyzing Half a Century of Debates and Data in Light of Speech Science," *Science Advances* 5, no. 12 (December 2019): eaaw3916.

38. Samir Okasha, "Biological Altruism," *The Stanford Encyclopedia of Philosophy* (Summer 2020 ed.), accessed November 29, 2022, https://plato.stanford.edu/archives /sum2020/entries/altruism-biological/.

39. Noam Chomsky, "Three Factors in Language Design," *Linguistic Inquiry* 36, no. 1 (2005): 3.

40. Cecilia S. L. Lai et al., "A Forkhead-Domain Gene Is Mutated in a Severe Speech and Language Disorder," *Nature* 413, no. 6855 (2001): 519–23.

41. Wolfgang Enard et al., "Molecular Evolution of *FOXP2*, a Gene Involved in Speech and Language," *Nature* 418, no. 6900 (2002): 869–72.

42. Elizabeth Grace Atkinson et al., "No Evidence for Recent Selection at *FOXP2* among Diverse Human Populations," *Cell* 174, no. 6 (2018): 1424–35.e15.

43. Johannes Krause et al., "The Derived *FOXP2* Variant of Modern Humans Was Shared with Neandertals," *Current Biology* 17, no. 21 (2007): 1908–12.

44. Bridget Alex, "Could Neanderthals Speak? The Ongoing Debate over Neanderthal Language," *Discover Magazine*, November 6, 2018, https://www.discovermagazine .com/planet-earth/could-neanderthals-speak-the-ongoing-debate-over-neanderthal -language.

45. Ray Jackendoff, "How Did Language Begin?" Linguistic Society of America, accessed October 20, 2022, https://www.linguisticsociety.org/resource/faq-how-did-language -begin.

46. Michael Tomasello et al., "Understanding and Sharing Intentions: The Origins of Cultural Cognition," *Behavioral and Brain Sciences* 28, no. 5 (2005): 687.

47. Tomasello et al., "Understanding and Sharing Intentions," 690.

48. Steven Pinker, "Language as an Adaptation to the Cognitive Niche," chap. 2 in *Language Evolution*, ed. Morten H. Christiansen and Simon Kirby (Oxford: Oxford University Press, 2003), 27.

49. Pinker, "Language as an Adaptation to the Cognitive Niche," 29.

50. Marc D. Hauser, Noam Chomsky, and W. Tecumseh Fitch, "The Faculty of Language: What Is It, Who Has It, and How Did It Evolve?" *Science* 298, no. 5598 (2002): 1569.

51. Hauser et al., "The Mystery of Language Evolution," 401.

52. Hauser, Chomsky, and Fitch, "The Faculty of Language," 1569–79.

53. Chris Knight, "Ritual/Speech Coevolution: A Solution to the Problem of Deception," chap. 5 in *Approaches to the Evolution of Language*, ed. James R. Hurford, Michael Studdert-Kennedy and Chris Knight (Cambridge: Cambridge University Press 1998), 74.

54. Roy A. Rappaport, *Ritual and Religion in the Making of Humanity* (Cambridge: Cambridge University Press, 1999), 15.

55. Chris Knight, "The Origins of Symbolic Culture," chap. 14 in *Homo Novus—A Human Without Illusions*, ed. Ulrich J. Frey, Charlotte Störmer and Kai P. Willführ (Berlin: Springer, 2010), 198.

56. Giuseppe Di Pellegrino et al., "Understanding Motor Events: A Neurophysiological Study," *Experimental Brain Research* 91, no. 1 (1992): 176–80; Vittorio Gallese et al., "Action Recognition in the Premotor Cortex," *Brain* 119, no. 2 (1996): 593–609; Giacomo Rizzolatti and Laila Craighero, "The Mirror-Neuron System," *Annual Review of Neuroscience* 27 (2004): 169–92; Giacomo Rizzolatti et al., "Premotor Cortex and the Recognition of Motor Actions," *Cognitive Brain Research* 3, no. 2 (1996): 131–41.

57. Evelyne Kohler et al., "Hearing Sounds, Understanding Actions: Action Representation in Mirror Neurons," *Science* 297, no. 5582 (2002): 846–48.

58. Michael A. Arbib and Mihail Bota, "Language Evolution: Neural Homologies and Neuroinformatics," *Neural Networks* 16, no. 9 (2003): 1247–52; Michael A. Arbib and Mihail Bota, "Neural Homologies and the Grounding of Neurolinguistics," ed. Michael A. Arbib, *Action to Language via the Mirror Neuron System* (Cambridge: Cambridge University Press, 2006), 152–54; Marco Iacoboni et al., "Cortical Mechanisms of Human Imitation," *Science* 286, no. 5449 (1999): 2526–28.

59. Michael Arbib and Giacomo Rizzolatti, "Neural Expectations: A Possible Evolutionary Path from Manual Skills to Language," in *The Nature of Concepts: Evolution, Structure and Representation*, ed. Philip Van Loocke (London: Routledge, 1998), 140–51; Michael A. Arbib, "The Mirror System Hypothesis on the Linkage of Action and Language," ed. Michael A. Arbib, *Action to Language via the Mirror Neuron System* (Cambridge: Cambridge University Press, 2006), 6–7; Giacomo Rizzolatti and Michael A. Arbib, "Language within our Grasp," *Trends in Neurosciences* 21, no. 5 (1998): 188–94; Giacomo Rizzolatti and Laila Craighero, "Language and Mirror Neurons," in *Oxford Handbook of Psycholinguistics*, ed. M. Gareth Gaskell (Oxford: Oxford University Press, 2007), 777–83.

60. Arbib and Rizzolatti, "Neural Expectations," 149–50.

61. Gordon W. Hewes et al., "Primate Communication and the Gestural Origin of Language," *Current Anthropology* 14, no. 1/2 (February–April 1973), 5–11; Michael C. Corballis, "The Gestural Origins of Language: Human Language May Have Evolved from Manual Gestures, Which Survive Today as a 'Behavioral Fossil' Coupled to Speech," *American Scientist* 87, no. 2 (1999): 139–43.

62. Arbib and Bota, "Neural Homologies and the Grounding of Neurolinguistics," 18–19.

63. Hauser et al., "The Mystery of Language Evolution," 401.

13. ON CREATIVITY

1. See Anna Abraham, *The Neuroscience of Creativity* (Cambridge: Cambridge University Press, 2018), and Rex E. Jung and Oshin Vartanian, eds., *The Cambridge Handbook of the Neuroscience of Creativity* (Cambridge: Cambridge University Press, 2018).

2. Abraham, *The Neuroscience of Creativity*, 9–11; Dean Keith Simonton, "Creative Ideas and the Creative Process: Good News and Bad News for the Neuroscience of Creativity," in *The Cambridge Handbook of the Neuroscience of Creativity*, ed. Rex E. Jung and Oshin Vartanian (Cambridge: Cambridge University Press, 2018), 9–10.

3. Melissa C. Duff et al., "Hippocampal Amnesia Disrupts Creative Thinking," *Hippocampus* 23, no. 12 (December 2013): 1143–49.

4. Arne Dietrich, "Types of Creativity," *Psychonomic Bulletin & Review* 26, no. 1 (February 2019): 1.

5. Dietrich, "Types of Creativity," 3.

6. Rex E. Jung et al., "The Structure of Creative Cognition in the Human Brain," *Frontiers in Human Neuroscience* 7 (2013): 330; Roger E. Beaty et al., "Creative Cognition and Brain Network Dynamics," *Trends in Cognitive Science* 20, no. 2 (February 2016): 88–93; Oshin Vartanian, "Neuroscience of Creativity," in *The Cambridge Handbook of Creativity*, ed. James C. Kaufman and Robert J. Sternberg (Cambridge: Cambridge University Press, 2019), 156–59.

7. Daniel L. Schacter and Donna Rose Addis, "The Cognitive Neuroscience of Constructive Memory: Remembering the Past and Imagining the Future," *Philosophical Transactions of the Royal Society B: Biological Sciences* 362, no. 1481 (2007): 773–75.

8. Marcela Ovando-Tellez et al., "Brain Connectivity-Based Prediction of Real-Life Creativity Is Mediated by Semantic Memory Structure," *Science Advances* 8, no. 5 (February 2022): eabl4294.

9. R. Keith Sawyer, *Explaining Creativity: The Science of Human Innovation* (New York: Oxford University Press, 2006), 153.

10. Chunfang Zhou and Lingling Luo, "Group Creativity in Learning Context: Understanding in a Social-Cultural Framework and Methodology," *Creative Education* 3, no. 4 (2012): 392; Amanda L. Thayer, Alexandra Petruzzelli, and Caitlin E. McClurg, "Addressing the Paradox of the Team Innovation Process: A Review and Practical Considerations," *American Psychologist* 73, no. 4 (2018): 363.

11. Rebecca Mitchell, Stephen Nicholas, and Brendan Boyle, "The Role of Openness to Cognitive Diversity and Group Processes in Knowledge Creation," *Small Group Research* 40, no. 5 (2009): 535–54; Zhou and Luo, "Group Creativity in Learning Context," 393; Paul B. Paulus, Jonali Baruah, and Jared B. Kenworthy, "Enhancing Collaborative Ideation in Organizations," *Frontiers in Psychology* 9 (2018): 2024.

12. Simonton, "Creative Ideas and the Creative Process," 15.

13. Alan J. Park et al., "Reset of Hippocampal-Prefrontal Circuitry Facilitates Learning," *Nature* 591, no. 7851 (March 2021): 615–19.

14. Mihaly Csikszentmihalyi, *Flow: The Psychology of Optimal Experience* (New York: HarperCollins, 2008), 71.

15. Lalit Kishore, "Kaon Parables for Awakening Intuitive Thinking in Zen Buddhism: An Example," Speakingtree.in, April 14, 2023, https://www.speakingtree.in/blog/kaon-parables-for-awakening-intuitive-thinking-in-zen-buddhism-an-example.

16. Jogye Order Missionary Office, "What Is Koan Contemplation Zen?" *Buddhist Newspaper* 2296, January 14, 2007, http://www.ibulgyo.com/news/articleView.html?idxno=78486.

17. Simonton, "Creative Ideas and the Creative Process," 16.

18. Yehuda Wacks and Aviv M. Weinstein, "Excessive Smartphone Use Is Associated with Health Problems in Adolescents and Young Adults," *Frontiers in Psychiatry* 12 (2021): 669042.

14. THE FUTURE OF INNOVATION

1. Hannah Ritchie and Max Roser, "Extinctions," Ourworldindata.org, accessed June 21, 2022, https://ourworldindata.org/extinctions.
2. "Living Planet Report 2022—Building a Nature-Positive Society," WWF.ca, October 12, 2022, https://wwf.ca/?s=Living+Planet+Report+2022&lang=en.
3. Michael Greshko and National Geographic Staff, "What Are Mass Extinctions, and What Causes Them?" *National Geographic*, September 26, 2019, https://www.national geographic.com/science/article/mass-extinction; "The 'Great Dying,'" Geological Society of America, May 19, 2021, www.sciencedaily.com/releases/2021/05/210519163702 .htm.
4. Hannah Ritchie and Max Roser, "Energy," Ourworldindata.org, accessed June 21, 2022, https://ourworldindata.org/energy; Gioietta Kuo, "When Fossil Fuels Run Out, What Then?" MAHB, May 23, 2019, https://mahb.stanford.edu/library-item/fossil-fuels-run/.
5. Dimitrios Floudas et al., "The Paleozoic Origin of Enzymatic Lignin Decomposition Reconstructed from 31 Fungal Genomes," *Science* 336, no. 6089 (June 2012): 1715–19.
6. As an alternative hypothesis, a unique combination of climate (in the tropical wet-lands, the massive amount of organic matter produced was slow to decay in the acidic water) and tectonics (the organic matter deposited by crustal thickening during the formation of Pangaea was preserved) has been proposed as the major factor for coal accumulation. Matthew P. Nelsen et al., "Delayed Fungal Evolution Did Not Cause the Paleozoic Peak in Coal Production," *Proceedings of the National Academy of Sciences of the United States* 113, no. 9 (March 2016): 2442–47.
7. Myles R. Allen et al., "Framing and Context," in *Global Warming of 1.5°C.*, ed. V. Masson-Delmotte et al. (Cambridge: Cambridge University Press, 2018), 51.
8. Bruce W. Sellwood and Paul J. Valdes, "Jurassic Climates," *Proceedings of the Geologists' Association* 119, no. 1 (2008): 5; Jessica E. Tierney et al., "Glacial Cooling and Climate Sensitivity Revisited," *Nature* 584, no. 7822 (2020): 569–73.
9. Paul J. Young et al., "The Montreal Protocol Protects the Terrestrial Carbon Sink," *Nature* 596, no. 7872 (2021): 384–88.
10. David Silver et al., "Reward Is Enough," *Artificial Intelligence* 299 (2021): 103535.
11. Ray Kurzweil, *The Singularity Is Near: When Humans Transcend Biology* (New York: Penguin, 2005), 135–36.
12. Irving John Good, "Speculations Concerning the First Ultraintelligent Machine," in *Advances in Computers* (Amsterdam: Elsevier, 1966), 33.
13. Kurzweil, *The Singularity Is Near*, 24.
14. Kurzweil, *The Singularity Is Near*, 28.
15. Maclyn McCarty, "Discovering Genes Are Made of DNA," *Nature* 421, no. 6921 (2003): 406.
16. Richard Evans and Jim Gao, "DeepMind AI Reduces Google Data Centre Cooling Bill by 40 Percent," DeepMind, July 6, 2016, https://www.deepmind.com/blog /deepmind-ai-reduces-google-data-centre-cooling-bill-by-40.
17. Precise durations are unclear; the estimated durations vary across studies.

EPILOGUE

1. Ray Kurzweil, *The Singularity Is Near: When Humans Transcend Biology* (New York: Penguin, 2005), 135–36.

APPENDIX 1: DENTATE GYRUS

1. Jong W. Lee and Min W. Jung, "Separation or Binding? Role of the Dentate Gyrus in Hippocampal Mnemonic Processing," *Neuroscience & Biobehavioral Reviews* 75 (April 2017): 184–91.
2. David Marr, "A Theory of Cerebellar Cortex," *Journal of Physiology* 202, no. 2 (June 1969): 440, 442–43; James S. Albus, "A Theory of Cerebellar Function," *Mathematical Biosciences* 10, nos. 1–2 (1971): 36–41.
3. David G. Amaral, Norio Ishizuka, and Brenda Claiborne, "Neurons, Numbers and the Hippocampal Network," *Progress in Brain Research* 83 (1990): 3.
4. Bruce L. McNaughton, "Neuronal Mechanisms for Spatial Computation and Information Storage," in *Neural Connections, Mental Computations*, ed. Lynn Nadel, Lynn A. Cooper, Peter W. Culicover and Robert M. Harnish (Cambridge, MA: MIT Press, 1989), 305; Edmund T. Rolls, "Functions of Neuronal Networks in the Hippocampus and Cerebral Cortex in Memory," in *Models of Brain Function*, ed. Rodney M. J. Cotterill (Cambridge: Cambridge University Press, 1989), 18–21; James J. Knierim and Joshua Neunuebel, "Tracking the Flow of Hippocampal Computation: Pattern Separation, Pattern Completion, and Attractor Dynamics," *Neurobiology of Learning and Memory* 129 (March 2016): 39–46.
5. James J. Knierim, Inah Lee, and Eric L. Hargreaves, "Hippocampal Place Cells: Parallel Input Streams, Subregional Processing, and Implications for Episodic Memory," *Hippocampus* 16, no. 9 (2006): 760–62.
6. Lee and Jung, "Separation or Binding?" 183–94.
7. Raymond P. Kesner, "A Behavioral Analysis of Dentate Gyrus Function," *Progress in Brain Research* 163 (2007): 567–68; Sen Cheng, "The CRISP Theory of Hippocampal Function in Episodic Memory," *Frontiers in Neural Circuits* 7 (2013): 88.
8. Lee and Jung, "Separation or Binding?" 183–94.
9. Christina Herold, Vincent J. Coppola, and Verner P. Bingman, "The Maturation of Research into the Avian Hippocampal Formation: Recent Discoveries from One of the Nature's Foremost Navigators," *Hippocampus* 25, no. 11 (November 2015): 1196; Georg F. Striedter, "Evolution of the Hippocampus in Reptiles and Birds," *Journal of Comparative Neurology* 524, no. 3 (February 2016): 505–6.
10. Verner P. Bingman and Rubén N. Muzio, "Reflections on the Structural-Functional Evolution of the Hippocampus: What Is the Big Deal about a Dentate Gyrus?" *Brain, Behavior and Evolution* 90, no. 1 (2017): 56–59.
11. Tim Bliss, Graham Collingridge, and Richard Morris, "Synaptic Plasticity in the Hippocampus," in *The Hippocampus Book*, ed. Per Andersen et al. (Oxford: Oxford University Press, 2007), 344–46.

APPENDIX 2: VALUE-CODING NEURONS

1. Hyunjung Lee et al., "Hippocampal Neural Correlates for Values of Experienced Events," *Journal of Neuroscience* 32, no. 43 (October 2012): 15062.
2. Eric B. Knudsen and Joni D. Wallis, "Hippocampal Neurons Construct a Map of an Abstract Value Space," *Cell* 184, no. 18 (September 2021): 4640–50.e10.
3. In many natural settings, potential competitors must also be considered when making a choice. A high-value target would have more competitors. Hence, always choosing the high-value target is not an optimal strategy in a social competition situation.

APPENDIX 2: VALUE-CODING NEURONS

Humans and animals have a natural tendency to distribute their choices in proportion to the probability of reward, which is called probability matching. It is considered an optimal strategy in the presence of competitors because they cannot exploit your choice behavior. Charles R. Gallistel, *The Organization of Learning* (Cambridge, MA: MIT Press, 1990), 351–53.

4. Richard S. Sutton and Andrew G. Barto, *Reinforcement Learning: An Introduction* (Cambridge, MA: MIT Press, 1998), 148–51.

5. Lee et al., "Hippocampal Neural Correlates for Values of Experienced Events," 15053–65.

BIBLIOGRAPHY

Abraham, Anna. *The Neuroscience of Creativity.* Cambridge: Cambridge University Press, 2018.

Addis, Donna R., Alana T. Wong, and Daniel L. Schacter. "Remembering the Past and Imagining the Future: Common and Distinct Neural Substrates During Event Construction and Elaboration." *Neuropsychologia* 45, no. 7 (April 2007): 1363–77.

Ai, Hiroyuki, Ryuichi Okada, Midori Sakura, Thomas Wachtler, and Hidetoshi Ikeno. "Neuroethology of the Waggle Dance: How Followers Interact with the Waggle Dancer and Detect Spatial Information." *Insects* 10, no. 10 (October 2019): 336.

Aiello, Leslie C., and Peter Wheeler. "The Expensive-Tissue Hypothesis: The Brain and the Digestive System in Human and Primate Evolution." *Current Anthropology* 36, no. 2 (1995): 199–221.

Albus, James S. "A Theory of Cerebellar Function." *Mathematical Biosciences* 10, nos. 1–2 (1971): 25–61.

Alex, Bridget. "Could Neanderthals Speak? The Ongoing Debate over Neanderthal Language." *Discover*, November 6, 2018, https://www.discovermagazine.com/planet-earth/could-neanderthals-speak-the-ongoing-debate-over-neanderthal-language.

Allen, Myles R., et al. "Framing and Context." In *Global Warming of 1.5°C*, ed. V. Masson-Delmotte et al., 49–92. Cambridge: Cambridge University Press, 2018.

Amaral, David G., Norio Ishizuka, and Brenda Claiborne. "Neurons, Numbers and the Hippocampal Network." *Progress in Brain Research* 83 (1990): 1–11.

Ambrose, R. Ellen, Brad E. Pfeiffer, and David J. Foster. "Reverse Replay of Hippocampal Place Cells Is Uniquely Modulated by Changing Reward." *Neuron* 91, no. 5 (September 2016): 1124–36.

Arbib, Michael A. "The Mirror System Hypothesis on the Linkage of Action and Language." In *Action to Language via the Mirror Neuron System*, ed. Michael A. Arbib, 2–48. Cambridge: Cambridge University Press, 2006.

Arbib, Michael A., and Mihail Bota. "Language Evolution: Neural Homologies and Neuroinformatics." *Neural Networks* 16, no. 9 (2003): 1237–60.

Arbib, Michael A., and Mihail Bota. "Neural Homologies and the Grounding of Neuro-linguistics." In *Action to Language via the Mirror Neuron System*, ed. Michael A. Arbib, 135–74. Cambridge: Cambridge University Press, 2006.

Arbib, Michael, and Giacomo Rizzolatti. "Neural Expectations: A Possible Evolutionary Path from Manual Skills to Language." In *The Nature of Concepts: Evolution, Structure and Representation*, ed. Philip Van Loocke, 128–54. London: Routledge, 1998.

Atkinson, Elizabeth Grace, Amanda Jane Audesse, Julia Adela Palacios, Dean Michael Bobo, Ashley Elizabeth Webb, Sohini Ramachandran, and Brenna Mariah Henn. "No Evidence for Recent Selection at *FOXP2* among Diverse Human Populations." *Cell* 174, no. 6 (2018): 1424–35.e15.

Badre, David. "Cognitive Control, Hierarchy, and the Rostro-Caudal Organization of the Frontal Lobes." *Trends in Cognitive Science* 12, no. 5 (May 2008): 193–200.

Baek, Seungdae, Min Song, Jaeson Jang, Gwangsu Kim, and Se-Bum Paik. "Face Detection in Untrained Deep Neural Networks." *Nature Communications* 12, no. 1 (December 2021): 7328.

Bar-Yosef, Ofer. "The Upper Paleolithic Revolution." *Annual Review of Anthropology* 31, no. 1 (2002): 363–93.

Beaty, Roger E., Mathias Benedek, Paul J. Silvia, and Daniel L. Schacter. "Creative Cognition and Brain Network Dynamics." *Trends in Cognitive Science* 20, no. 2 (February 2016): 87–95.

Beaty, Roger E. "The Creative Brain." *Cerebrum* (January 2020): cer-02-20.

Beaudet, Amélie, Andrew Du, and Bernard Wood. "Evolution of the Modern Human Brain." *Progress in Brain Research* 250 (2019): 219–50.

Beecher, Michael D. "Why Are No Animal Communication Systems Simple Languages?" *Frontiers in Psychology* (2021): 701.

Behrens, Timothy E. J., Timothy H. Muller, James C. R. Whittington, Shirley Mark, Alon B. Baram, Kimberly L. Stachenfeld, and Zeb Kurth-Nelson. "What Is a Cognitive Map? Organizing Knowledge for Flexible Behavior." *Neuron* 100, no. 2 (October 2018): 490–509.

Ben-Yishay, Elhanan, Ksenia Krivoruchko, Shaked Ron, Nachum Ulanovsky, Dori Derdikman, and Yoram Gutfreund. "Directional Tuning in the Hippocampal Formation of Birds." *Current Biology* 31, no. 12 (June 2021): 2592–602.e4.

Bhattarai, Baburam, Jong W. Lee, and Min W. Jung. "Distinct Effects of Reward and Navigation History on Hippocampal Forward and Reverse Replays." *Proceedings of the National Academy of Sciences of the United States of America* 117, no. 1 (January 2020): 689–97.

Bingman, Verner P., and Rubén N. Muzio. "Reflections on the Structural-Functional Evolution of the Hippocampus: What Is the Big Deal about a Dentate Gyrus?" *Brain, Behavior and Evolution* 90, no. 1 (2017): 53–61.

Bingman, Verner P., Cosme Salas, and Fernando Rodríguez. "Evolution of the Hippocampus." In *Encyclopaedia of Neuroscience*, ed. Marc D. Binder, Nobutaka Hirokawa, and Uwe Windhorst, 1356–60. Berlin: Springer, 2009.

Bissonette, Gregory B., and Matthew R. Roesch. "Neural Correlates of Rules and Conflict in Medial Prefrontal Cortex During Decision and Feedback Epochs." *Frontiers in Behavioral Neuroscience* 9 (2015): 266.

Bliss, T. V., and A. R. Gardner-Medwin. "Long-Lasting Potentiation of Synaptic Transmission in the Dentate Area of the Unanaesthetized Rabbit Following Stimulation of the Perforant Path." *Journal of Physiology* 232, no. 2 (July 1973): 357–74.

Bliss, T. V., and T. Lomo. "Long-Lasting Potentiation of Synaptic Transmission in the Dentate Area of the Anaesthetized Rabbit Following Stimulation of the Perforant Path." *Journal of Physiology* 232, no. 2 (July 1973): 331–56.

Bliss, Tim, Graham Collingridge, and Richard Morris. "Synaptic Plasticity in the Hippocampus." Chap. 10 in *The Hippocampus Book*, ed. Per Andersen, Richard Morris, David Amaral, Tim Bliss, and John O'Keefe, 343–474. Oxford: Oxford University Press, 2007.

Bloom, Paul. "Language and Thought: Does Grammar Makes Us Smart?" *Current Biology* 10, no. 14 (2000): R516–R17.

Bloom, Paul, and Frank C. Keil. "Thinking through Language." *Mind & Language* 16, no. 4 (2001): 351–67.

Boë, Louis-Jean, Thomas R. Sawallis, Joël Fagot, Pierre Badin, Guillaume Barbier, Guillaume Captier, Lucie Ménard, Jean-Louis Heim, and Jean-Luc Schwartz. "Which Way to the Dawn of Speech? Reanalyzing Half a Century of Debates and Data in Light of Speech Science." *Science Advances* 5, no. 12 (December 2019): eaaw3916.

Borghi, Anna M., Laura Barca, Ferdinand Binkofski, Cristiano Castelfranchi, Giovanni Pezzulo, and Luca Tummolini. "Words as Social Tools: Language, Sociality and Inner Grounding in Abstract Concepts." *Physics of Life Reviews* 29 (2019): 120–53.

Bruce, Darryl. "Fifty Years Since Lashley's 'In Search of the Engram': Refutations and Conjectures." *Journal of the History of the Neurosciences* 10, no. 3 (2001): 308–18.

Bruner, Emiliano. "Human Paleoneurology and the Evolution of the Parietal Cortex." *Brain, Behavior and Evolution* 91, no. 3 (2018): 136–47.

Bruner, Emiliano, Hideki Amano, José M. de la Cuétara, and Naomichi Ogihara. "The Brain and the Braincase: A Spatial Analysis on the Midsagittal Profile in Adult Humans." *Journal of Anatomy* 227, no. 3 (September 2015): 268–76.

Bruner, Emiliano, and Sofia Pereira-Pedro. "A Metric Survey on the Sagittal and Coronal Morphology of the Precuneus in Adult Humans." *Brain Structure and Function* 225, no. 9 (December 2020): 2747–55.

Bruner, Emiliano, Todd M. Preuss, Xu Chen, and James K. Rilling. "Evidence for Expansion of the Precuneus in Human Evolution." *Brain Structure and Function* 222, no. 2 (March 2017): 1053–60.

Bruner, Emiliano, Francisco J. Roman, José M. de la Cuétara, Manue Martin-Loeches, and Roberto Colom. "Cortical Surface Area and Cortical Thickness in the Precuneus of Adult Humans." *Neuroscience* 286 (February 2015): 345–52.

Buckner, Randy L., Jessica R. Andrews-Hanna, and Daniel L. Schacter. "The Brain's Default Network: Anatomy, Function, and Relevance to Disease." *Annals of the New York Academy of Sciences* 1124 (March 2008): 1–38.

Bulganin, Lisa, and Bianca C. Wittmann. "Reward and Novelty Enhance Imagination of Future Events in a Motivational-Episodic Network." *PLoS One* 10, no. 11 (2015): e0143477.

Buzsaki, Gyorgy. "Hippocampal Sharp Wave-Ripple: A Cognitive Biomarker for Episodic Memory and Planning." *Hippocampus* 25, no. 10 (October 2015): 1073–1188.

Bynum, William F. "Wernicke, Carl." Encyclopedia.com. Updated May 8, 2018. https://www.encyclopedia.com/people/medicine/psychology-and-psychiatry-biographies/carl-wernicke.

Cann, Rebecca L., Mark Stoneking, and Allan C. Wilson. "Mitochondrial DNA and Human Evolution." *Nature* 325, no. 6099 (1987): 31–36.Carmody, Rachel N., and Richard W. Wrangham. "The Energetic Significance of Cooking." *Journal of Human Evolution* 57, no. 4 (October 2009): 379–91.

Cavanna, Andrea E., and Michael R. Trimble. "The Precuneus: A Review of Its Functional Anatomy and Behavioural Correlates." *Brain* 129, part 3 (March 2006): 564–83.

Chamberlain, Rebecca, I. Chris McManus, Nicola Brunswick, Qona Rankin, Howard Riley, and Ryota Kanai. "Drawing on the Right Side of the Brain: A Voxel-Based Morphometry Analysis of Observational Drawing." *Neuroimage* 96 (August 2014): 167–73.

Chen, Qun-Lin, Ting Xu, Wen-Jing Yang, Ya-Dan Li, Jiang-Zhou Sun, Kang-Cheng Wang, Roger E. Beaty et al. "Individual Differences in Verbal Creative Thinking Are Reflected in the Precuneus." *Neuropsychologia* 75 (August 2015): 441–49.

Chen, Yvonne Y., Lyndsey Aponik-Gremillion, Eleonora Bartoli, Daniel Yoshor, Sameer A. Sheth, and Brett L. Foster. "Stability of Ripple Events During Task Engagement in Human Hippocampus." *Cell Reports* 35, no. 13 (2021): 109304.

Cheng, Qi, Austin Roth, Eric Halgren, and Rachel I. Mayberry. "Effects of Early Language Deprivation on Brain Connectivity: Language Pathways in Deaf Native and Late First-Language Learners of American Sign Language." *Frontiers in Human Neuroscience* 13 (2019): 320.

Cheng, Sen. "The CRISP Theory of Hippocampal Function in Episodic Memory." *Frontiers in Neural Circuits* 7 (2013): 88.

Choi, Sang-Han, Gangwon Jeong, Young-Bo Kim, and Zang-Hee Cho. "Proposal for Human Visual Pathway in the Extrastriate Cortex by Fiber Tracking Method Using Diffusion-Weighted MRI." *Neuroimage* 220 (October 2020): 117145.

Chomsky, Noam. "Three Factors in Language Design." *Linguistic Inquiry* 36, no. 1 (2005): 1–22.

Constantinescu, Alexandra O., Jill X. O'Reilly, and Timothy E. J. Behrens. "Organizing Conceptual Knowledge in Humans with a Gridlike Code." *Science* 352, no. 6292 (June 2016): 1464–68.

Corballis, Michael C. "The Gestural Origins of Language: Human Language May Have Evolved from Manual Gestures, Which Survive Today as a 'Behavioral Fossil' Coupled to Speech." *American Scientist* 87, no. 2 (1999): 138–45.

Corkin, Suzanne. "What's New with the Amnesic Patient H. M.?" *Nature Reviews Neuroscience* 3, no. 2 (February 2002): 153–60.

Cornelio, Alianda M., Ruben E. de Bittencourt-Navarrete, Ricardo de Bittencourt Brum, Claudio M. Queiroz, and Marcos R. Costa. "Human Brain Expansion During Evolution Is Independent of Fire Control and Cooking." *Frontiers in Neuroscience* 10 (2016): 167.

Craig, Michael. "Memory and Forgetting." In *Encyclopedia of Behavioral Neuroscience*, ed. Sergio Della Sala, 425–31 (Amsterdam: Elsevier, 2021).

Csikszentmihalyi, Mihaly. *Flow: The Psychology of Optimal Experience.* New York: HarperCollins, 2008.

Davies, Stephen. "Behavioral Modernity in Retrospect." *Topoi* 40, no. 1 (2021): 221–32.

Dennis, Patricia, Stephen M. Shuster, and C. N. Slobodchikoff. "Dialects in the Alarm Calls of Black-Tailed Prairie Dogs (*Cynomys ludovicianus*): A Case of Cultural Diffusion?" *Behavioral Processes* 181 (December 2020): 104243.

Denzel, Stephanie. "George Franklin." The National Registry of Exonerations. Updated May 2, 2022, https://www.law.umich.edu/special/exoneration/Pages/casedetail.aspx?caseid=3221.

Di Pellegrino, Giuseppe, Luciano Fadiga, Leonardo Fogassi, Vittorio Gallese, and Giacomo Rizzolatti. "Understanding Motor Events: A Neurophysiological Study." *Experimental Brain Research* 91, no. 1 (1992): 176–80.

Diamond, Jared. "The Great Leap Forward." In *Technology and Society: Issue for the 21st Century and Beyond*, 3rd ed., ed. Linda Hjorth, Barbara Eichler, Ahmed Khan and John Morello, 15–24. Upper Saddle River, NJ: Prentice Hall, 2008.

Diba, Kamran, and Gyorgy Buzsaki. "Forward and Reverse Hippocampal Place-Cell Sequences During Ripples." *Nature Neuroscience* 10, no. 10 (October 2007): 1241–42.

DiCarlo, James J., Davide Zoccolan, and Nicole C. Rust. "How Does the Brain Solve Visual Object Recognition?" *Neuron* 73, no. 3 (February 2012): 415–34.

Dietrich, Arne. "Types of Creativity." *Psychonomic Bulletin & Review* 26, no. 1 (February 2019): 1–12.

Donahue, Chad J., Matthew F. Glasser, Todd M. Preuss, James K. Rilling, and David C. Van Essen. "Quantitative Assessment of Prefrontal Cortex in Humans Relative to Nonhuman Primates." *Proceedings of the National Academy of Sciences of the United States of America* 115, no. 22 (May 2018): E5183–E92.

Dove, Guy. "More Than a Scaffold: Language Is a Neuroenhancement." *Cognitive Neuropsychology* 37, no. 5–6 (2020): 288–311.

Dronkers, Nina F., Odile Plaisant, Marie Therese Iba-Zizen, and Emmanuel A. Cabanis. "Paul Broca's Historic Cases: High Resolution MR Imaging of the Brains of Leborgne and Lelong." *Brain* 130, no. 5 (2007): 1432–41.

Duff, Melissa C., Jake Kurczek, Rachael Rubin, Neal J. Cohen, and Daniel Tranel. "Hippocampal Amnesia Disrupts Creative Thinking." *Hippocampus* 23, no. 12 (December 2013): 1143–9.

Dumontheil, Iroise. "Development of Abstract Thinking During Childhood and Adolescence: The Role of Rostrolateral Prefrontal Cortex." *Developmental Cognitive Neuroscience* 10 (October 2014): 57–76.

Duncan, Katherine, Bradley B. Doll, Nathaniel D. Daw, and Daphna Shohamy. "More Than the Sum of Its Parts: A Role for the Hippocampus in Configural Reinforcement Learning." *Neuron* 98, no. 3 (May 2018): 645–57.

Dupret, David, Joseph O'Neill, Barty Pleydell-Bouverie, and Jozsef Csicsvari. "The Reorganization and Reactivation of Hippocampal Maps Predict Spatial Memory Performance." *Nature Neuroscience* 13, no. 8 (August 2010): 995–1002.

Eichenbaum, Howard, Paul Dudchenko, Emma Wood, Matthew Shapiro, and Heikki Tanila. "The Hippocampus, Memory, and Place Cells: Is It Spatial Memory or a Memory Space?" *Neuron* 23, no. 2 (June 1999): 209–26.

Ekstrom, Arne D., Michael J. Kahana, Jeremy B. Caplan, Tony A. Fields, Eve A. Isham, Ehren L. Newman, and Itzhak Fried. "Cellular Networks Underlying Human Spatial Navigation." *Nature* 425, no. 6954 (September 2003): 184–8.

Enard, Wolfgang, Molly Przeworski, Simon E. Fisher, Cecilia S. L. Lai, Victor Wiebe, Takashi Kitano, Anthony P. Monaco, and Svante Pääbo. "Molecular Evolution of *FOXP2*, a Gene Involved in Speech and Language." *Nature* 418, no. 6900 (2002): 869–72.

Entomological Society of America. "Frequently Asked Questions on Entomology." Updated July 26, 2010, https://www.entsoc.org/resources/faq/.

Evans, Richard, and Jim Gao. "DeepMind AI Reduces Google Data Centre Cooling Bill by 40 Percent." DeepMind, July 6, 2016, https://www.deepmind.com/blog/deepmind -ai-reduces-google-data-centre-cooling-bill-by-40.

Floudas, Dimitrios, Manfred Binder, Robert Riley, Kerrie Barry, Robert A. Blanchette, Bernard Henrissat, Angel T. Martinez, et al. "The Paleozoic Origin of Enzymatic Lignin Decomposition Reconstructed from 31 Fungal Genomes." *Science* 336, no. 6089 (June 2012): 1715–19.

Fonseca-Azevedo, Karina, and Suzana Herculano-Houzel. "Metabolic Constraint Imposes Tradeoff between Body Size and Number of Brain Neurons in Human Evolution." *Proceedings of the National Academy of Sciences of the United States of America* 109, no. 45 (November 2012): 18571–6.

Foster, David J., and Matthew A. Wilson. "Reverse Replay of Behavioural Sequences in Hippocampal Place Cells During the Awake State." *Nature* 440, no. 7084 (March 2006): 680–83.

Gallese, Vittorio, Luciano Fadiga, Leonardo Fogassi, and Giacomo Rizzolatti. "Action Recognition in the Premotor Cortex." *Brain* 119, no. 2 (1996): 593–609.

Gallistel, Charles R. *The Organization of Learning.* Cambridge, MA: MIT Press, 1990.

Gangopadhyay, Prabaha, Megha Chawla, Olga Dal Monte, and Steve W. C. Chang. "Prefrontal-Amygdala Circuits in Social Decision-Making." *Nature Neuroscience* 24, no. 1 (January 2023): 5–18.

Geological Society of America. "The 'Great Dying.'" May 19, 2021, www.sciencedaily.com /releases/2021/05/210519163702.htm.

Geva-Sagiv, Maya, Liora Las, Yossi Yovel, and Nachum Ulanovsky. "Spatial Cognition in Bats and Rats: From Sensory Acquisition to Multiscale Maps and Navigation." *Nature Reviews Neuroscience* 16, no. 2 (February 2015): 94–108.

Gleitman, Lila, and Anna Papafragou. "Language and Thought." In *Cambridge Handbook of Thinking and Reasoning*, ed. Keith Holyoak and Robert G. Morrison, 633–61. Cambridge: Cambridge University Press, 2005.

Glickman, Neil, Charlene Crump, and Steve Hamerdinger. "Language Deprivation Is a Game Changer for the Clinical Specialty of Deaf Mental Health." *JADARA* 54, no. 1 (2020): 54–89.

Good, Irving John. "Speculations Concerning the First Ultraintelligent Machine." In *Advances in Computers*, 31–88. Amsterdam: Elsevier, 1966.

Greshko, Michael, and National Geographic Staff. "What Are Mass Extinctions, and What Causes Them?" *National Geographic*, September 26, 2019, https://www.national geographic.com/science/article/mass-extinction.

Gruber, Matthias J., Maureen Ritchey, Shao-Fang Wang, Manoj K. Doss, and Charan Ranganath. "Post-Learning Hippocampal Dynamics Promote Preferential Retention of Rewarding Events." *Neuron* 89, no. 5 (March 2016): 1110–20.

Gulati, Sanjay. "Language Deprivation Syndrome." In *Language Deprivation and Deaf Mental Health*, ed. Neil S. Glickman and Wyatte C. Hall, 24–53. London: Routledge, 2018.

Gunz, Philipp, Amanda K. Tilot, Katharina Wittfeld, Alexander Teumer, Chin Y. Shapland, Theo G. M. van Erp, Michael Dannemann, et al. "Neandertal Introgression Sheds Light on Modern Human Endocranial Globularity." *Current Biology* 29, no. 1 (January 2019): 120–27.

Gupta, Anoopum S., Matthijs A. A. van der Meer, David S. Touretzky, and A. David Redish. "Hippocampal Replay Is Not a Simple Function of Experience." *Neuron* 65, no. 5 (March 2010): 695–705.

Gusnard, Debra A., and Marcus E. Raichle. "Searching for a Baseline: Functional Imaging and the Resting Human Brain." *Nature Reviews Neuroscience* 2, no. 10 (October 2001): 685–94.

Hafting, Torkel, Marianne Fyhn, Sturla Molden, May-Britt Moser, and Edvard I. Moser. "Microstructure of a Spatial Map in the Entorhinal Cortex." *Nature* 436, no. 7052 (August 2005): 801–6.

Hagoort, Peter. "MUC (Memory, Unification, Control) and Beyond." *Frontiers in Psychology* 4 (2013): 416.

Hall, Matthew L., Inge-Marie Eigsti, Heather Bortfeld, and Diane Lillo-Martin. "Auditory Deprivation Does Not Impair Executive Function, but Language Deprivation Might: Evidence from a Parent-Report Measure in Deaf Native Signing Children." *The Journal of Deaf Studies and Deaf Education* 22, no. 1 (2017): 9–21.

Hall, Wyatte C., Leonard L. Levin, and Melissa L. Anderson. "Language Deprivation Syndrome: A Possible Neurodevelopmental Disorder with Sociocultural Origins." *Social Psychiatry and Psychiatric Epidemiology* 52, no. 6 (2017): 761–76.

Harari, Yuval Noah. *Sapiens: A Brief History of Humankind*. Trans. Hyun-Wook Cho. Paju, South Korea: Gimm-Young, 2015.

Hargreaves, Eric L., Geeta Rao, Inah Lee, and James J. Knierim. "Major Dissociation between Medial and Lateral Entorhinal Input to Dorsal Hippocampus." *Science* 308, no. 5729 (June 2005): 1792–94.

Harlow, John M. "Recovery from the Passage of an Iron Bar through the Head." *Publication of the Massachusetts Medical Society* 2, no. 3 (1868): 329–47.

Hassabis, Demis, Dharshan Kumaran, Seralynne D. Vann, and Eleanor A. Maguire. "Patients with Hippocampal Amnesia Cannot Imagine New Experiences." *Proceedings of the National Academy of Sciences of the United States of America* 104, no. 5 (January 2007): 1726–31.

Hauser, Marc D., Noam Chomsky, and W. Tecumseh Fitch. "The Faculty of Language: What Is It, Who Has It, and How Did It Evolve?" *Science* 298, no. 5598 (2002): 1569–79.

Hauser, Marc D., Charles Yang, Robert C. Berwick, Ian Tattersall, Michael J. Ryan, Jeffrey Watumull, Noam Chomsky, and Richard C. Lewontin. "The Mystery of Language Evolution." *Frontiers in Psychology* (2014): 401.

Hebb, Donald Olding. *The Organization of Behavior: A Psychological Theory*. New York: Wiley, 1949.

Heide, Michael, Christiane Haffner, Ayako Murayama, Yoko Kurotaki, Haruka Shinohara, Hideyuki Okano, Erika Sasaki, and Wieland B. Huttner. "Human-Specific *ARHGAP11B* Increases Size and Folding of Primate Neocortex in the Fetal Marmoset." *Science* 369, no. 6503 (July 2020): 546–50.

Hewes, Gordon W. "Primate Communication and the Gestural Origin of Language." *Current Anthropology* 14, no. 1/2 (February–April 1973), 5–24.

Herculano-Houzel, Suzana. "The Human Brain in Numbers: A Linearly Scaled-Up Primate Brain." *Frontiers in Human Neuroscience* 3 (2009): 31.

Herculano-Houzel, Suzana, Kamilla Avelino-de-Souza, Kleber Neves, Jairo Porfirio, Debora Messeder, Larissa Mattos Feijo, Jose Maldonado, and Paul R. Manger. "The Elephant Brain in Numbers." *Frontiers in Neuroanatomy* 8 (2014): 46.

Herold, Christina, Vincent J. Coppola, and Verner P. Bingman. "The Maturation of Research into the Avian Hippocampal Formation: Recent Discoveries from One of the Nature's Foremost Navigators." *Hippocampus* 25, no. 11 (November 2015): 1193–211.

Hetherington, Renée, and Robert G. B. Reid. *The Climate Connection: Climate Change and Modern Human Evolution*. Cambridge: Cambridge University Press, 2010.

Higgins, Cameron, Yunzhe Liu, Diego Vidaurre, Zeb Kurth-Nelson, Ray Dolan, Timothy Behrens, and Mark Woolrich. "Replay Bursts in Humans Coincide with Activation of the Default Mode and Parietal Alpha Networks." *Neuron* 109, no. 5 (March 2021): 882–93.e7.

Hockett, Charles F. "Animal "Languages" and Human Language." *Human Biology* 31, no. 1 (1959): 32–39.

Hockett, Charles F., and Charles, D. Hockett. "The Origin of Speech." *Scientific American* 203, no. 3 (1960): 88–97.

Hofman, Michel A. "Evolution of the Human Brain: When Bigger Is Better." *Frontiers in Neuroanatomy* 8 (2014): 15.

Howe, Mark L., and Lauren M. Knott. "The Fallibility of Memory in Judicial Processes: Lessons from the Past and Their Modern Consequences." *Memory* 23, no. 5 (2015): 633–56. Hublin, Jean-Jacques, Abdelouahed Ben-Ncer, Shara E. Bailey, Sarah E. Freidline, Simon Neubauer, Matthew M. Skinner, Inga Bergmann, et al. "New Fossils from Jebel Irhoud, Morocco and the Pan-African Origin of *Homo Sapiens*." *Nature* 546, no. 7657 (June 2017): 289–92.

Hussein, Basel A.-S. "The Sapir-Whorf Hypothesis Today." *Theory and Practice in Language Studies* 2, no. 3 (March 2012): 642–46.

Hyman, James M., Liya Ma, Emili Balaguer-Ballester, Daniel Durstewitz, and Jeremy K. Seamans. "Contextual Encoding by Ensembles of Medial Prefrontal Cortex Neurons." *Proceedings of the National Academy of Sciences of the United States of America* 109, no. 13 (March 2012): 5086–91.

Iacoboni, Marco, Roger P. Woods, Marcel Brass, Harold Bekkering, John C. Mazziotta, and Giacomo Rizzolatti. "Cortical Mechanisms of Human Imitation." *Science* 286, no. 5449 (1999): 2526–28.

Ivry, Richard B., and Juliana V. Baldo. "Is the Cerebellum Involved in Learning and Cognition?" *Current Opinion in Neurobiology* 2, no. 2 (1992): 212–16.

Jabr, Ferris. "How Humans Evolved Supersize Brains." *Quanta Magazine*, November 10, 2015, https://www.quantamagazine.org/how-humans-evolved-supersize-brains-20151110/.

Jackendoff, Ray. "How Did Language Begin?" Linguistic Society of America. Accessed October 20, 2022, https://www.linguisticsociety.org/resource/faq-how-did-language-begin.

Jauk, Emanuel, Aljoscha C. Neubauer, Beate Dunst, Andreas Fink, and Mathias Benedek. "Gray Matter Correlates of Creative Potential: A Latent Variable Voxel-Based Morphometry Study." *Neuroimage* 111 (May 1 2015): 312–20.

Jay, Thérèse M., Jacques Glowinski, and Anne-Marie Thierry. "Selectivity of the Hippo-campal Projection to the Prelimbic Area of the Prefrontal Cortex in the Rat." *Brain Research* 505, no. 2 (December 1989): 337–40.

Jeong, Yeongseok, Namjung Huh, Joonyeup Lee, Injae Yun, Jong W. Lee, Inah Lee, and Min W. Jung. "Role of the Hippocampal CA1 Region in Incremental Value Learning." *Scientific Reports* 8, no. 1 (June 2018): 9870.

Ji, Daoyun, and Matthew A. Wilson. "Coordinated Memory Replay in the Visual Cortex and Hippocampus During Sleep." *Nature Neuroscience* 10, no. 1 (January 2007): 100–7.

Jogye Order Missionary Office. "What Is Koan Contemplation Zen?" *Buddhist Newspaper* 2296, January 2007, http://www.ibulgyo.com/news/articleView.html?idxno=78486.

Joseph, R. Gabriel. *Frontal Lobes: Neuroscience, Neuropsychology, Neuropsychiatry* Cambridge, MA: Cosmology, 2011.

Jung, Min W., Hyunjung Lee, Yeongseok Jeong, Jong W. Lee, and Inah Lee. "Remem-bering Rewarding Futures: A Simulation-Selection Model of the Hippocampus." *Hippocampus* 28, no. 12 (December 2018): 913–30.

Jung, Rex E., Brittany S. Mead, Jessica Carrasco, and Ranee A. Flores. "The Structure of Creative Cognition in the Human Brain." *Frontiers in Human Neuroscience* 7 (2013): 330.

Jung, Rex E, and Oshin Vartanian. *The Cambridge Handbook of the Neuroscience of Creativity.* Cambridge: Cambridge University Press, 2018.

Kaas, Jon H. "The Origin and Evolution of Neocortex: From Early Mammals to Modern Humans." *Progress in Brain Research* 250 (2019): 61–81.

Kaas, Jon H., and Mary K. L. Baldwin. "The Evolution of the Pulvinar Complex in Primates and Its Role in the Dorsal and Ventral Streams of Cortical Processing." *Vision* 4, no. 1 (December 2019): 3.

Kant, Immanuel. *Critique of Pure Reason.* Ed. Paul Guyer and Allen W. Wood. Cambridge: Cambridge University Press, 1999.

Kesner, Raymond P., and John C. Churchwell. "An Analysis of Rat Prefrontal Cortex in Mediating Executive Function." *Neurobiology of Learning and Memory* 96, no. 3 (October 2011): 417–31.

Kesner, Raymond P. "A Behavioral Analysis of Dentate Gyrus Function." *Progress in Brain Research* 163 (2007): 567–76.

Kim, Gwangsu, Jaeson Jang, Seungdae Baek, Min Song, and Se-Bum Paik. "Visual Number Sense in Untrained Deep Neural Networks." *Science Advances* 7, no. 1 (2021): eabd6127.

King, Maedbh, Carlos R. Hernandez-Castillo, Russell A. Poldrack, Richard B. Ivry, and Jörn Diedrichsen. "Functional Boundaries in the Human Cerebellum Revealed by a Multi-Domain Task Battery." *Nature Neuroscience* 22, no. 8 (2019): 1371–78.

Kishore, Lalit. "Kaon Parables for Awakening Intuitive Thinking in Zen Buddhism: An Example." Speakingtree.in, April 14, 2023, https://www.speakingtree.in/blog/kaon -parables-for-awakening-intuitive-thinking-in-zen-buddhism-an-example.

Klein, Richard G. "Out of Africa and the Evolution of Human Behavior." *Evolutionary Anthropology: Issues, News, and Reviews* 17, no. 6 (2008): 267–81.

Knierim, James J., Inah Lee, and Eric L. Hargreaves. "Hippocampal Place Cells: Parallel Input Streams, Subregional Processing, and Implications for Episodic Memory." *Hippocampus* 16, no. 9 (2006): 755–64.

Knierim, James J., and Joshua Neunuebel. "Tracking the Flow of Hippocampal Computation: Pattern Separation, Pattern Completion, and Attractor Dynamics." *Neurobiology of Learning and Memory* 129 (March 2016): 38–49.

Knight, Chris. "The Origins of Symbolic Culture." Chap. 14 in *Homo Novus—A Human Without Illusions,* ed. Ulrich J. Frey, Charlotte Störmer and Kai P. Willführ, 193–211. Berlin: Springer, 2010.

——. "Ritual/Speech Coevolution: A Solution to the Problem of Deception." Chap. 5 in *Approaches to the Evolution of Language,* ed. James R. Hurford, Michael Studdert-Kennedy and Chris Knight, 68–91. Cambridge: Cambridge University Press, 1998.

Knudsen, Eric B., and Joni D. Wallis. "Hippocampal Neurons Construct a Map of an Abstract Value Space." *Cell* 184, no. 18 (September 2021): 4640–50.e10.

Kochiyama, Takanori, Naomichi Ogihara, Hiroki C. Tanabe, Osamu Kondo, Hideki Amano, Kunihiro Hasegawa, Hiromasa Suzuki, et al. "Reconstructing the Neanderthal Brain Using Computational Anatomy." *Scientific Reports* 8, no. 1 (April 2018): 6296.

Kohler, Evelyne, Christian Keysers, M. Alessandra Umilta, Leonardo Fogassi, Vittorio Gallese, and Giacomo Rizzolatti. "Hearing Sounds, Understanding Actions: Action Representation in Mirror Neurons." *Science* 297, no. 5582 (2002): 846–48.

Konnikova, Maria. "The Man Who Couldn't Speak and How He Revolutionized Psychology." *Scientific American,* February 3, 2018, https://blogs.scientificamerican .com/literally-psyched/the-man-who-couldnt-speakand-how-he-revolutionized -psychology/.

Krause, Johannes, Carles Lalueza-Fox, Ludovic Orlando, Wolfgang Enard, Richard E. Green, Hernán A. Burbano, Jean-Jacques Hublin, et al. "The Derived *FOXP2* Variant of Modern Humans Was Shared with Neandertals." *Current Biology* 17, no. 21 (2007): 1908–12.

Krupenye, Christopher, and Josep Call. "Theory of Mind in Animals: Current and Future Directions." *Wiley Interdisciplinary Reviews: Cognitive Science* 10, no. 6 (November 2019): e1503.

Kuo, Gioietta. "When Fossil Fuels Run out, What Then?" MAHB, May 23, 2019, https://mahb.stanford.edu/library-item/fossil-fuels-run/.

Kurth-Nelson, Zeb, Marcos Economides, Raymond J. Dolan, and Peter Dayan. "Fast Sequences of Non-Spatial State Representations in Humans." *Neuron* 91, no. 1 (July 6 2016): 194–204.

Kurzweil, Ray. *The Singularity Is Near: When Humans Transcend Biology*. New York: Penguin, 2005.

Lage, Caio A., De Wet Wolmarans, and Daniel C. Mograbi. "An Evolutionary View of Self-Awareness." *Behavioural Processes* 194 (January 2022): 104543.

Lai, Cecilia S. L., Simon E. Fisher, Jane A. Hurst, Faraneh Vargha-Khadem, and Anthony P. Monaco. "A Forkhead-Domain Gene Is Mutated in a Severe Speech and Language Disorder." *Nature* 413, no. 6855 (2001): 519–23.

Lanczik, Mario, and G. Keil. "Carl Wernicke's Localization Theory and Its Significance for the Development of Scientific Psychiatry." *History of Psychiatry* 2, no. 6 (1991): 171–80.

Langland-Hassan, Peter, Frank R. Faries, Maxwell Gatyas, Aimee Dietz, and Michael J. Richardson. "Assessing Abstract Thought and Its Relation to Language with a New Nonverbal Paradigm: Evidence from Aphasia." *Cognition* 211 (2021): 104622.

Langland-Hassan, Peter, Christopher Gauker, Michael J. Richardson, Aimee Dietz, and Frank R. Faries. "Metacognitive Deficits in Categorization Tasks in a Population with Impaired Inner Speech." *Acta Psychologica* 181 (2017): 62–74.

Le, Quoc V., Marc'Aurelio Ranzato, Rajat Monga, Matthieu Devin, Kai Chen, Greg S. Corrado, Jeff Dean, and Andrew Y. Ng. "Building High-Level Features Using Large Scale Unsupervised Learning." Paper presented at the Proceedings of the 29th International Conference on Machine Learning, 2012. https://doi.org/10.48550/arXiv.1112.6209.

LeCun, Yann, Yoshua Bengio, and Geoffrey Hinton. "Deep Learning." *Nature* 521, no. 7553 (May 2015): 436–44.

Lee, Albert K., and Matthew A. Wilson. "Memory of Sequential Experience in the Hippocampus During Slow Wave Sleep." *Neuron* 36, no. 6 (December 2002): 1183–94.

Lee, Daeyeol, Hyojung Seo, and Min W. Jung. "Neural Basis of Reinforcement Learning and Decision Making." *Annual Review of Neuroscience* 35 (2012): 287–308.

Lee, Hyunjung, Jeong-Wook Ghim, Hoseok Kim, Daeyeol Lee, and Min W. Jung. "Hippocampal Neural Correlates for Values of Experienced Events." *Journal of Neuroscience* 32, no. 43 (October 2012): 15053–65.

Lee, Jong W., and Min W. Jung. "Separation or Binding? Role of the Dentate Gyrus in Hippocampal Mnemonic Processing." *Neuroscience & Biobehavioral Reviews* 75 (April 2017): 183–94.

Lee, Sung-Hyun, Namjung Huh, Jong W. Lee, Jeong-Wook Ghim, Inah Lee, and Min W. Jung. "Neural Signals Related to Outcome Evaluation Are Stronger in CA1 Than CA3." *Frontiers in Neural Circuits* 11 (2017): 40.

Lehr, Andrew B., Arvind Kumar, Christian Tetzlaff, Torkel Hafting, Marianne Fyhn, and Tristan M. Stober. "CA2 Beyond Social Memory: Evidence for a Fundamental Role in Hippocampal Information Processing." *Neuroscience & Biobehavioral Reviews* 126 (July 2021): 398–412.

Levy, William B. "A Sequence Predicting CA3 Is a Flexible Associator That Learns and Uses Context to Solve Hippocampal-Like Tasks." *Hippocampus* 6, no. 6 (1996): 579–90.

Liu, Anli A., Simon Henin, Saman Abbaspoor, Anatol Bragin, Elizabeth A. Buffalo, Jordan S. Farrell, David J. Foster, et al. "A Consensus Statement on Detection of Hippocampal Sharp Wave Ripples and Differentiation from Other Fast Oscillations." *Nature Communications* 13 (2022): 6000.

Liu, Yunzhe, Raymond J. Dolan, Zeb Kurth-Nelson, and Timothy E. J. Behrens. "Human Replay Spontaneously Reorganizes Experience." *Cell* 178, no. 3 (July 2019): 640–52.e14.

Loftus, Elizabeth F., and Katherine Ketcham. "Truth or Invention: Exploring the Repressed Memory Syndrome; Excerpt from 'The Myth of Repressed Memory.'" *Cosmopolitan*, April 1995, https://staff.washington.edu/eloftus/Articles/Cosmo.html.

Loftus, Elizabeth F. "Planting Misinformation in the Human Mind: A 30-Year Investigation of the Malleability of Memory." *Learning & Memory* 12, no. 4 (2005): 361–66.

Loftus, Elizabeth F., and Jacqueline E. Pickrell. "The Formation of False Memories." *Psychiatric Annals* 25, no. 12 (December 1995): 720–25.

Logothetis, Nikos K., Oxana Eschenko, Yusuke Murayama, Mark Augath, Thomas Steudel, Henry C. Evrard, Michel Besserve, and Axel Oeltermann. "Hippocampal-Cortical Interaction During Periods of Subcortical Silence." *Nature* 491, no. 7425 (November 2012): 547–53.

Louie, Kenway, and Matthew A. Wilson. "Temporally Structured Replay of Awake Hippocampal Ensemble Activity During Rapid Eye Movement Sleep." *Neuron* 29, no. 1 (January 2001): 145–56. Lundervold, Alexander S., and Arvid Lundervold. "An Overview of Deep Learning in Medical Imaging Focusing on MRI." *Zeitschrift für Medizinische Physik* 29, no. 2 (May 2019): 102–27.

Lupyan, Gary, and Bodo Winter. "Language Is More Abstract Than You Think, or, Why Aren't Languages More Iconic?" *Philosophical Transactions of the Royal Society B: Biological Sciences* 373, no. 1752 (2018): 20170137.

Malik, Ruchi, Yi Li, Selin Schamiloglu, and Vikaas S. Sohal. "Top-Down Control of Hippocampal Signal-to-Noise by Prefrontal Long-Range Inhibition." *Cell* 185, no. 9 (April 2022): 1602–17.

Manns, Joseph R., and Howard Eichenbaum. "Evolution of Declarative Memory." *Hippocampus* 16, no. 9 (2006): 795–808.

Mansouri, Farshad A., David J. Freedman, and Mark J. Buckley. "Emergence of Abstract Rules in the Primate Brain." *Nature Reviews Neuroscience* 21, no. 11 (November 2020): 595–610.

Marr, David. "A Theory of Cerebellar Cortex." *Journal of Physiology* 202, no. 2 (June 1969): 437–70.

——. "Simple Memory: A Theory for Archicortex." *Philosophical Transactions of the Royal Society B: Biological Sciences* 262, no. 841 (July 1971): 23–81.

Mayberry, Rachel I., Jen-Kai Chen, Pamela Witcher, and Denise Klein. "Age of Acquisition Effects on the Functional Organization of Language in the Adult Brain." *Brain and Language* 119, no. 1 (2011): 16–29.

McCarty, Maclyn. "Discovering Genes Are Made of DNA." *Nature* 421, no. 6921 (2003): 406.

McClelland, James L., Bruce L. McNaughton, and Randall C. O'Reilly. "Why There Are Complementary Learning Systems in the Hippocampus and Neocortex: Insights from the Successes and Failures of Connectionist Models of Learning and Memory." *Psychology Review* 102, no. 3 (July 1995): 419–57. McCormick, Cornelia, Clive R. Rosenthal, Thomas D. Miller, and Eleanor A. Maguire. "Mind-Wandering in People with Hippocampal Damage." *Journal of Neuroscience* 38, no. 11 (March 2018): 2745–54. McDonald, Robert J., and Norman M. White. "A Triple Dissociation of Memory Systems: Hippocampus, Amygdala, and Dorsal Striatum." *Behavioral Neuroscience* 107, no. 1 (February 1993): 3–22. McDougall, Ian, Francis H. Brown, and John G. Fleagle. "Stratigraphic Placement and Age of Modern Humans from Kibish, Ethiopia." *Nature* 433, no. 7027 (2005): 733–36.

McNaughton, Bruce L. "Neuronal Mechanisms for Spatial Computation and Information Storage." In *Neural Connections, Mental Computations*, ed. Lynn Nadel, Lynn A. Cooper, Peter W. Culicover and Robert M. Harnish, 285–350. Cambridge, MA: MIT Press, 1989.

Melton, Buckner F., Jr. "George Franklin Trial: 1990–91." Encyclopedia.com. Accessed June 28, 2022, https://www.encyclopedia.com/law/law-magazines/george-franklin -trial-1990-91.

Menzel, Randolf. "The Waggle Dance as an Intended Flight: A Cognitive Perspective." *Insects* 10, no. 12 (November 2019): 424.

Mitchell, Rebecca, Stephen Nicholas, and Brendan Boyle. "The Role of Openness to Cognitive Diversity and Group Processes in Knowledge Creation." *Small Group Research* 40, no. 5 (2009): 535–54.

Morriss-Kay, Gillian M. "The Evolution of Human Artistic Creativity." *Journal of Anatomy* 216, no. 2 (February 2010): 158–76.

Murphy, Patrick, Tim Shallice, Gail Robinson, Sarah E. MacPherson, Martha Turner, Katherine Woollett, Marco Bozzali, and Lisa Cipolotti. "Impairments in Proverb Interpretation Following Focal Frontal Lobe Lesions." *Neuropsychologia* 51, no. 11 (September 2013): 2075–86.

Nadel, Lynn, Alexei Samsonovich, Lee Ryan, and Morris Moscovitch. "Multiple Trace Theory of Human Memory: Computational, Neuroimaging, and Neuropsychological Results." *Hippocampus* 10, no. 4 (2000): 352–68.

Nelsen, Matthew P., William A. DiMichele, Shanan E. Peters, and C. Kevin Boyce. "Delayed Fungal Evolution Did Not Cause the Paleozoic Peak in Coal Production." *Proceedings of the National Academy of Sciences of the United States of America* 113, no. 9 (March 2016): 2442–47.

Neubauer, Simon, Jean-Jacques Hublin, and Philipp Gunz. "The Evolution of Modern Human Brain Shape." *Science Advances* 4, no. 1 (2018): eaao5961.

News Staff. "Breakthrough of the Year: The Runners-Up." *Science* 318, no. 5858 (December 2007): 1844–49.

Nieder, Andreas. "Supramodal Numerosity Selectivity of Neurons in Primate Prefrontal and Posterior Parietal Cortices." *Proceedings of the National Academy of Sciences of the United States of America* 109, no. 29 (July 2012): 11860–5.

——. "The Neuronal Code for Number." *Nature Reviews Neuroscience* 17, no. 6 (2016): 366–82.

Nobelförsamlingen. "Press Release. The Nobel Prize in Physiology or Medicine 2014," October 6, 2014, https://www.nobelprize.org/prizes/medicine/2014/press-release/.

O'Connell, Sanjida. "The Perils of Relying on Memory in Court." *Telegraph*, December 15, 2008, https://www.telegraph.co.uk/technology/3778272/The-perils-of-relying-on-memory-in-court.html.

O'Keefe, John, and Jonathan Dostrovsky. "The Hippocampus as a Spatial Map: Preliminary Evidence from Unit Activity in the Freely-Moving Rat." *Brain Research* 34, no. 1 (November 1971): 171–5.

O'Keefe, John, and Lynn Nadel. *The Hippocampus as a Cognitive Map*. Oxford: Clarendon, 1978.

Ofshe, Richard J. "Inadvertent Hypnosis During Interrogation: False Confession Due to Dissociative State; Mis-identified Multiple Personality and the Satanic Cult Hypothesis." *International Journal of Clinical and Experimental Hypnosis* 40, no. 3 (1992): 125–56.

Okasha, Samir. "Biological Altruism." *The Stanford Encyclopedia of Philosophy* (Summer 2020 ed.), accessed November 29, 2022, https://plato.stanford.edu/archives/sum2020/entries/altruism-biological/.

Olafsdottir, H. Freyja, Caswell Barry, Aman B. Saleem, Demis Hassabis, and Hugo J. Spiers. "Hippocampal Place Cells Construct Reward Related Sequences through Unexplored Space." *eLife* 4 (June 2015): e06063. Olio, Karen A., and William F. Cornell. "The Facade of Scientific Documentation: A Case Study of Richard Ofshe's Analysis of the Paul Ingram Case." *Psychology, Public Policy, and Law* 4, no. 4 (1998): 1182–97.

Ovando-Tellez, Marcela, Yoed N. Kenett, Mathias Benedek, Matthieu Bernard, Joan Belo, Benoit Beranger, Theophile Bieth, and Emmanuelle Volle. "Brain Connectivity-Based Prediction of Real-Life Creativity Is Mediated by Semantic Memory Structure." *Science Advances* 8, no. 5 (February 4 2022): eabl4294.

Packard, Mark G., and Barbara J. Knowlton. "Learning and Memory Functions of the Basal Ganglia." *Annual Review of Neuroscience* 25 (2002): 563–93.

Padoa-Schioppa, Camillo, and Katherine E. Conen. "Orbitofrontal Cortex: A Neural Circuit for Economic Decisions." *Neuron* 96, no. 4 (November 2017): 736–54.

Pagel, Mark. "Q&A: What Is Human Language, When Did It Evolve and Why Should We Care?" *BMC Biology* 15, no. 1 (July 2017): 64.

Park, Alan J., Alexander Z. Harris, Kelly M. Martyniuk, Chia-Yuan Chang, Atheir I. Abbas, Daniel C. Lowes, Christoph Kellendonk, Joseph A. Gogos, and Joshua A. Gordon. "Reset of Hippocampal-Prefrontal Circuitry Facilitates Learning." *Nature* 591, no. 7851 (March 2021): 615–19.

Park, Seongmin A., Douglas S. Miller, and Erie D. Boorman. "Inferences on a Multidimensional Social Hierarchy Use a Grid-Like Code." *Nature Neuroscience* 24, no. 9 (September 2021): 1292–301.

Parloff, Roger. "Why Deep Learning Is Suddenly Changing Your Life." *Fortune*, September 29, 2016, https://fortune.com/longform/ai-artificial-intelligence-deep-machine-learning/.

Patterson, Francine, and Wendy Gordon. "The Case for the Personhood of Gorillas." In *The Great Ape Project*, ed. Paola Cavalieri and Peter Singer, 58–77. New York: St. Martin's, 1993.

Patterson, Francine, and Wendy Gordon. "Twenty-Seven Years of Project Koko and Michael." In *All Apes Great and Small*, 165–76. Berlin: Springer, 2002.

Patzke, Nina, Muhammad A. Spocter, Karl Æ. Karlsson, Mads F. Bertelsen, Mark Haagensen, Richard Chawana, Sonja Streicher, et al. "In Contrast to Many Other Mammals, Cetaceans Have Relatively Small Hippocampi That Appear to Lack Adult Neurogenesis." *Brain Structure and Function* 220, no. 1 (January 2015): 361–83.

Paulus, Paul B., Jonali Baruah, and Jared B. Kenworthy, "Enhancing Collaborative Ideation in Organizations." *Frontiers in Psychology* 9 (2018): 2024.

Payne, Hannah L., Galen F. Lynch, and Dmitriy Aronov. "Neural Representations of Space in the Hippocampus of a Food-Caching Bird." *Science* 373, no. 6552 (July 2021): 343–48.

Perlovsky, Leonid. "Language and Cognition—Joint Acquisition, Dual Hierarchy, and Emotional Prosody." *Frontiers in Behavioral Neuroscience* 7 (2013): 123.

Pezzulo, Giovanni, Matthijs A. A. van der Meer, Carien S. Lansink, and Cyriel M. A. Pennartz. "Internally Generated Sequences in Learning and Executing Goal-Directed Behavior." *Trends in Cognitive Sciences* 18, no. 12 (December 2014): 647–57.

Pfeiffer, Brad E., and David J. Foster. "Hippocampal Place-Cell Sequences Depict Future Paths to Remembered Goals." *Nature* 497, no. 7447 (May 2013): 74–79.

Pinker, Steven. "Language as an Adaptation to the Cognitive Niche." Chap. 2 in *Language Evolution*, ed. Morten H. Christiansen and Simon Kirby, 16–37. Oxford: Oxford University Press, 2003.

Poeppel, David, Karen Emmorey, Gregory Hickok, and Liina Pylkkanen. "Towards a New Neurobiology of Language." *Journal of Neuroscience* 32, no. 41 (October 2012): 14125–31.

Powell, Adam, Stephen Shennan, and Mark G. Thomas. "Late Pleistocene Demography and the Appearance of Modern Human Behavior." *Science* 324, no. 5932 (June 2009): 1298–301.

Pullum, Geoffrey K. "Koko Is Dead, but the Myth of Her Linguistic Skills Lives On." The Chronicle of Higher Education, June 27, 2018, https://www.chronicle.com/blogs/linguafranca/koko-is-dead-but-the-myth-of-her-linguistic-skills-lives-on.

Quiroga, Rodrigo Q. "Concept Cells: The Building Blocks of Declarative Memory Functions." *Nature Reviews Neuroscience* 13, no. 8 (July 2012): 587–97.

Quiroga, Rodrigo Q., Leila Reddy, Gabriel Kreiman, Christof Koch, and Itzhak Fried. "Invariant Visual Representation by Single Neurons in the Human Brain." *Nature* 435, no. 7045 (June 2005): 1102–7.

Raichle, Marcus E., Ann Mary MacLeod, Abraham Z. Snyder, William J. Powers, Debra A. Gusnard, and Gordon L. Shulman. "A Default Mode of Brain Function." *Proceedings of the National Academy of Sciences of the United States of America* 98, no. 2 (January 2001): 676–82.

Rajasethupathy, Priyamvada, Sethuraman Sankaran, James H. Marshel, Christina K. Kim, Emily Ferenczi, Soo Y. Lee, Andre Berndt, et al. "Projections from Neocortex Mediate Top-Down Control of Memory Retrieval." *Nature* 526, no. 7575 (October 2015): 653–9.

Ramirez, Steve, Xu Liu, Pei-Ann Lin, Junghyup Suh, Michele Pignatelli, Roger L. Redondo, Tomás J. Ryan, et al. "Creating a False Memory in the Hippocampus." *Science* 341, no. 6144 (July 2013): 387–91.

Rappaport, Roy A. *Ritual and Religion in the Making of Humanity*. Cambridge: Cambridge University Press, 1999.

Reinert, Sandra, Mark Hübener, Tobias Bonhoeffer, and Pieter M. Goltstein. "Mouse Prefrontal Cortex Represents Learned Rules for Categorization." *Nature* 593, no. 7859 (May 2021): 411–17.

Rescorla, Michael. "The Language of Thought Hypothesis." The Stanford Encyclopedia of Philosophy (Summer 2019 Edition). Accessed November 11, 2022, https://plato.stanford.edu/archives/sum2019/entries/language-thought/.

Rey, Hernan G., Belen Gori, Fernando J. Chaure, Santiago Collavini, Alejandro O. Blenkmann, Pablo Seoane, Eduardo Seoane, Silvia Kochen, and Rodrigo Q. Quiroga. "Single Neuron Coding of Identity in the Human Hippocampal Formation." *Current Biology* 30, no. 6 (March 2020): 1152–59 e3. Rich, Erin L., and Matthew Shapiro. "Rat Prefrontal Cortical Neurons Selectively Code Strategy Switches." *Journal of Neuroscience* 29, no. 22 (June 2009): 7208–19.

Riel-Salvatore, Julien, and Claudine Gravel-Miguel. "Upper Palaeolithic Mortuary Practices in Eurasia: A Critical Look at the Burial Record." In *The Oxford Handbook of the Archaeology of Death and Burial*, ed. Liv N. Stutz and Sarah Tarlow, 303–46. Oxford: Oxford University Press, 2013.

Ritchie, Hannah, and Max Roser. "Energy." Ourworldindata.org. Accessed June 21, 2022, https://ourworldindata.org/energy.

——. "Extinctions." Ourworldindata.org. Accessed June 21, 2022, https://ourworldindata .org/extinctions.

Rizzolatti, Giacomo, and Michael A. Arbib. "Language within Our Grasp." *Trends in Neurosciences* 21, no. 5 (1998): 188–94. https://doi.org/10.1016/S0166-2236(98)01260-0.

Rizzolatti, Giacomo, and Laila Craighero. "Language and Mirror Neurons." In *Oxford Handbook of Psycholinguistics*, ed. M. Gareth Gaskell, 771–85. Oxford: Oxford University Press, 2007.

——. "The Mirror-Neuron System." *Annual Review of Neuroscience* 27 (2004): 169–92. https://doi.org/10.1146/annurev.neuro.27.070203.144230.

Rizzolatti, Giacomo, Luciano Fadiga, Vittorio Gallese, and Leonardo Fogassi. "Premotor Cortex and the Recognition of Motor Actions." *Cognitive Brain Research* 3, no. 2 (1996): 131–41.

Roberts, Reece P., and Donna Rose Addis. "A Common Mode of Processing Governing Divergent Thinking and Future Imagination." In *The Cambridge Handbook of the Neuroscience of Creativity*, ed. Rex E. Jung and Oshin Vartanian, 211–30. Cambridge: Cambridge University Press, 2018.

Roediger, Henry L. and Kathleen B. McDermott. "Creating False Memories: Remembering Words Not Presented in Lists." *Journal of Experimental Psychology: Learning, Memory, and Cognition* 21, no. 4 (1995): 803–14.

Rolls, Edmund T. "Functions of Neuronal Networks in the Hippocampus and Cerebral Cortex in Memory." In *Models of Brain Function*, ed. Rodney M. J. Cotterill, 15–33. Cambridge: Cambridge University Press, 1989.

Royal Entomological Society. "Facts and Figures." Understanding Insects, n.d., https:// www.royensoc.co.uk/facts-and-figures.

Russell, Bertrand. *Human Knowledge: Its Scope and Limits*. London: Routledge, 2009.

Ryle, Gilbert. *The Concept of Mind*. New York: Barnes and Noble, 1949.

Samhita, Laasya, and Hans J. Gross. "The "Clever Hans Phenomenon" Revisited." *Communicative & Integrative Biology* 6, no. 6 (November 2013): e27122.

Sawyer, R. Keith. *Explaining Creativity: The Science of Human Innovation*. New York: Oxford University Press, 2006.

Schacter, Daniel L. "Constructive Memory: Past and Future." *Dialogues in Clinical Neuroscience* 14, no. 1 (March 2012): 7–18.

Schacter, Daniel L., and Donna Rose Addis. "The Cognitive Neuroscience of Constructive Memory: Remembering the Past and Imagining the Future." *Philosophical Transactions of the Royal Society B: Biological Sciences* 362, no. 1481 (2007): 773–86.

Schlaggar, Bradley L., and Dennis D. O'Leary. "Potential of Visual Cortex to Develop an Array of Functional Units Unique to Somatosensory Cortex." *Science* 252, no. 5012 (June 1991): 1556–60.

Schuck, Nicolas W., and Yael Niv. "Sequential Replay of Nonspatial Task States in the Human Hippocampus." *Science* 364, no. 6447 (June 2019).

Schultz, Wolfram, Paul Apicella, Tomas Ljungberg, Ranulfo Romo, and Eugenio Scarnati. "Reward-Related Activity in the Monkey Striatum and Substantia Nigra." *Progress in Brain Research* 99 (1993): 227–35.

Schultz, Wolfram, Peter Dayan, and P. Read Montague. "A Neural Substrate of Prediction and Reward." *Science* 275, no. 5306 (March 1997): 1593–99.

Schulz, Hannes, and Sven Behnke. "Deep Learning: Layer-Wise Learning of Feature Hierarchies." *Künstliche Intelligenz* 26, no. 4 (2012): 357–63.

Scoville, William B., and Brenda Milner. "Loss of Recent Memory After Bilateral Hippocampal Lesions." *Journal of Neurology and Neurosurgical Psychiatry* 20, no. 1 (February 1957): 11–21.

Seger, Carol A., and Earl K. Miller. "Category Learning in the Brain." *Annual Review of Neuroscience* 33 (2010): 203–19.

Sellwood, Bruce W., and Paul J. Valdes. "Jurassic Climates." *Proceedings of the Geologists' Association* 119, no. 1 (2008): 5–17.

Shallice, Tim, and Lisa Cipolotti. "The Prefrontal Cortex and Neurological Impairments of Active Thought." *Annual Review of Psychology* 69 (January 2018): 157–80.

Shettleworth, Sara J. "Spatial Memory in Food-Storing Birds." *Philosophical Transactions of the Royal Society B: Biological Sciences* 329, no. 1253 (1990): 143–51.

Shettleworth, Sara J. *Cognition, Evolution, and Behavior*, 2nd ed. Oxford: Oxford University Press, 2010.

Shin, Eun J., Yunsil Jang, Soyoun Kim, Hoseok Kim, Xinying Cai, Hyunjung Lee, Jung H. Sul, et al. "Robust and Distributed Neural Representation of Action Values." *eLife* 10 (April 2021): e53045.

Siegal, Michael, and Rosemary Varley. "Aphasia, Language, and Theory of Mind." *Social Neuroscience* 1, no. 3–4 (2006): 167–74.

Siegel, Jennifer J., Douglas Nitz, and Verner P. Bingman. "Spatial-Specificity of Single-Units in the Hippocampal Formation of Freely Moving Homing Pigeons." *Hippocampus* 15, no. 1 (2005): 26–40.

——. "Lateralized Functional Components of Spatial Cognition in the Avian Hippocampal Formation: Evidence from Single-Unit Recordings in Freely Moving Homing Pigeons." *Hippocampus* 16, no. 2 (2006): 125–40.

Siegle, Joshua H., Xiaoxuan Jia, Séverine Durand, Sam Gale, Corbett Bennett, Nile Graddis, Greggory Heller, *et al.* "Survey of Spiking in the Mouse Visual System Reveals Functional Hierarchy." *Nature* 592, no. 7852 (April 2021): 86–92.

Silver, David, Satinder Singh, Doina Precup, and Richard S. Sutton. "Reward Is Enough." *Artificial Intelligence* 299 (2021): 103535.

Simonton, Dean Keith. "Creative Ideas and the Creative Process: Good News and Bad News for the Neuroscience of Creativity." In *The Cambridge Handbook of the Neuroscience of Creativity*, ed. Rex E. Jung and Oshin Vartanian, 9–18. Cambridge: Cambridge University Press, 2018.

Singer, Annabelle C., and Loren M. Frank. "Rewarded Outcomes Enhance Reactivation of Experience in the Hippocampus." *Neuron* 64, no. 6 (December 2009): 910–21.

Skaggs, William E., and Bruce L. McNaughton. "Replay of Neuronal Firing Sequences in Rat Hippocampus During Sleep Following Spatial Experience." *Science* 271, no. 5257 (March 1996): 1870–73.

Slater, Lauren. *Opening Skinner's Box: Great Psychological Experiments of the Twentieth Century.* New York: Norton, 2005.

Sliwa, Julia, Aurélie Planté, Jean-René Duhamel, and Sylvia Wirth. "Independent Neuronal Representation of Facial and Vocal Identity in the Monkey Hippocampus and Inferotemporal Cortex." *Cereb Cortex* 26, no. 3 (March 2016): 950–66.

Slobodchikoff, Con N., Andrea Paseka, and Jennifer L. Verdolin. "Prairie Dog Alarm Calls Encode Labels about Predator Colors." *Animal Cognition* 12, no. 3 (May 2009): 435–39.

Smithsonian. "Numbers of Insects (Species and Individuals)," Buginfo, Information Sheet Number 18, 1996, https://www.si.edu/spotlight/buginfo/bugnos.

Squire, Larry R. "The Legacy of Patient H. M. for Neuroscience." *Neuron* 61, no. 1 (January 2019): 6. Squire, Larry R., and Pablo Alvarez. "Retrograde Amnesia and Memory Consolidation: A Neurobiological Perspective." *Current Opinions in Neurobiology* 5, no. 2 (April 1995): 169–77. Stella, Federico, Peter Baracskay, Joseph O'Neill, and Jozsef Csicsvari. "Hippocampal Reactivation of Random Trajectories Resembling Brownian Diffusion." *Neuron* 102, no. 2 (April 2019): 450–61 e7.

Stevens, T. A., and J. R. Krebs. "Retrieval of Stored Seeds by Marsh Tits *Parus Palustris* in the Field." *Ibis* 128, no. 4 (1986): 513–25.

Stork, Nigel E. "How Many Species of Insects and Other Terrestrial Arthropods Are There on Earth?" *Annual Review of Entomology* 63 (2018): 31–45.

Striedter, Georg F. "Evolution of the Hippocampus in Reptiles and Birds." *Journal of Comparative Neurology* 524, no. 3 (February 2016): 496–517.

Sutton, Richard S., and Andrew G. Barto. *Reinforcement Learning: An Introduction.* Cambridge, MA: MIT Press, 1998.

Sutton, Richard S. "Dyna, an Integrated Architecture for Learning, Planning, and Reacting." *ACM Sigart Bulletin* 2, no. 4 (1991): 160–63.

Szpunar, Karl K., Jason M. Watson, and Kathleen B. McDermott. "Neural Substrates of Envisioning the Future." *Proceedings of the National Academy of Sciences of the United States of America* 104, no. 2 (January 2007): 642–7.

Tait, David S., E. Alexander Chase, and Verity J. Brown. "Attentional Set-Shifting in Rodents: A Review of Behavioural Methods and Pharmacological Results." *Current Pharmaceutical Design* 20, no. 31 (2014): 5046–59.

Tanaka, Saori C., Kenji Doya, Go Okada, Kazutaka Ueda, Yasumasa Okamoto, and Shigeto Yamawaki. "Prediction of Immediate and Future Rewards Differentially Recruits Cortico-Basal Ganglia Loops." *Nature Neuroscience* 7, no. 8 (August 2004): 887–93.

Tanaka, Shoji, and Eiji Kirino. "Reorganization of the Thalamocortical Network in Musicians." *Brain Research* 1664 (June 2017): 48–54.

Thayer, Amanda L., Alexandra Petruzzelli, and Caitlin E. McClurg. "Addressing the Paradox of the Team Innovation Process: A Review and Practical Considerations." *American Psychologist* 73, no. 4 (2018): 363–375.

Thurman, Judith. "First Impressions: What Does the World's Oldest Art Say about Us?" *New Yorker*, June 16, 2008, https://www.newyorker.com/magazine/2008/06/23/first -impressions.

BIBLIOGRAPHY

Tierney, Jessica E., Jiang Zhu, Jonathan King, Steven B. Malevich, Gregory J. Hakim, and Christopher J. Poulsen. "Glacial Cooling and Climate Sensitivity Revisited." *Nature* 584, no. 7822 (2020): 569–73.

Timmermann, Axel. "Quantifying the Potential Causes of Neanderthal Extinction: Abrupt Climate Change Versus Competition and Interbreeding." *Quaternary Science Reviews* 238 (2020).

Tomasello, Michael, Malinda Carpenter, Josep Call, Tanya Behne, and Henrike Moll. "Understanding and Sharing Intentions: The Origins of Cultural Cognition." *Behavioral and Brain Sciences* 28, no. 5 (2005): 675–91.

Tremblay, Sébastien, K. M. Sharika, and Michael L. Platt. "Social Decision-Making and the Brain: A Comparative Perspective." *Trends in Cognitive Sciences* 21, no. 4 (April 2017): 265–76. Tzakis, Nikolaos, and Matthew R. Holahan. "Social Memory and the Role of the Hippocampal CA2 Region." *Frontiers in Behavioral Neuroscience* 13 (2019): 233.

Utevsky, Amanda V., David V. Smith, and Scott A. Huettel. "Precuneus Is a Functional Core of the Default-Mode Network." *Journal of Neuroscience* 34, no. 3 (January 2014): 932–40.

Valladas, Hélène, Jean-Louis Reyss, Jean-Louis Joron, Georges Valladas, Ofer Bar-Yosef, and Bernard Vandermeersch. "Thermoluminescence Dating of Mousterian Troto-Cro-Magnon Remains from Israel and the Origin of Modern Man." *Nature* 331, no. 6157 (1988): 614–16.

van der Meer, Matthijs, Zeb Kurth-Nelson, and A. David Redish. "Information Processing in Decision-Making Systems." *Neuroscientist* 18, no. 4 (August 2012): 342–59.

Van Essen, David C., Chad J. Donahue, and Matthew F. Glasser. "Development and Evolution of Cerebral and Cerebellar Cortex." *Brain, Behavior and Evolution* 91, no. 3 (2018): 158–69.

Varley, Rosemary A., Nicolai J. C. Klessinger, Charles A. J. Romanowski, and Michael Siegal. "Agrammatic but Numerate." *Proceedings of the National Academy of Sciences of the United States of America* 102, no. 9 (2005): 3519–24.

Varley, Rosemary, and Michael Siegal. "Evidence for Cognition without Grammar from Causal Reasoning and 'Theory of Mind' in an Agrammatic Aphasic Patient." *Current Biology* 10, no. 12 (2000): 723–26.

Vartanian, Oshin. "Neuroscience of Creativity." In *The Cambridge Handbook of Creativity*, ed. James C. Kaufman and Robert J. Sternberg, 148–72. Cambridge: Cambridge University Press, 2019.

Wacewicz, Sławomir, and Przemysław Żywiczyński. "Language Evolution: Why Hockett's Design Features Are a Non-Starter." *Biosemiotics* 8, no. 1 (2015): 29–46.

Wacks, Yehuda, and Aviv M. Weinstein. "Excessive Smartphone Use Is Associated with Health Problems in Adolescents and Young Adults." *Frontiers in Psychiatry* 12 (2021): 669042.

Wade, Kimberley A., Maryanne Garry, J. Don Read, and D. Stephen Lindsay. "A Picture Is Worth a Thousand Lies: Using False Photographs to Create False Childhood Memories." *Psychonomic Bulletin & Review* 9, no. 3 (September 2002): 597–603.

Wallace, Douglas C., Michael D. Brown, and Marie T. Lott. "Mitochondrial DNA Variation in Human Evolution and Disease." *Gene* 238, no. 1 (September 1999): 211–30.

Wikenheiser, Andrew M., and Geoffrey Schoenbaum. "Over the River, through the Woods: Cognitive Maps in the Hippocampus and Orbitofrontal Cortex." *Nature Reviews Neuroscience* 17, no. 8 (August 2016): 513–23.

Wikipedia. "Gilbert Ryle." Updated October, 2021, https://en.wikipedia.org/wiki/Gilbert
_Ryle.
——. "Rodney Halbower." Updated April 27, 2022, https://en.wikipedia.org/wiki/Rodney
_Halbower.
——. "Thurston County Ritual Abuse Case." Updated December, 2021, https://en.wikipedia
.org/wiki/Thurston_County_ritual_abuse_case.
Wilde, Oscar. "The Critic as Artist." Accessed November 11, 2022, https://celt.ucc.ie
/published/E800003-007/text001.html.
Wiley, R. Haven. "Design Features of Language." In *Encyclopedia of Evolutionary Psycho-
logical Science*, 1919–31. Berlin: Springer, 2021.
Willoughby, Pamela R. "Modern Human Behavior." In *Oxford Research Encyclopedia
of Anthropology*. Published online May 29, 2020, https://doi.org/10.1093/acrefore
/9780190854584.013.46.
Wittgenstein, Ludwig. *Tractatus Logico-Philosophicus*. Trans. C. K. Ogden. London:
Routledge, 1981.
Wright, Lawrence. "Remembering Satan—Part II. What Was Going On in Thurston
County?" *New Yorker*, May 16, 1993, https://www.newyorker.com/magazine/1993/05/24
/remembering-satan-part-ii.
WWF. "Living Planet Report 2022—Building a Nature-Positive Society." WWF.ca,
October 12, 2022, https://wwf.ca/?s=Living+Planet+Report+2022&lang=en.
Xing, Lei, Agnieszka Kubik-Zahorodna, Takashi Namba, Anneline Pinson, Marta Florio,
Jan Prochazka, Mihail Sarov, Radislav Sedlacek, and Wieland B. Huttner. "Expres-
sion of Human-Specific *ARHGAP11B* in Mice Leads to Neocortex Expansion and
Increased Memory Flexibility." *EMBO Journal* 40, no. 13 (July 2021): e107093.
Yang, Sunggu, Sungchil Yang, Thais Moreira, Gloria Hoffman, Greg C. Carlson, Kevin
J. Bender, Bradley E. Alger, and Cha-Min Tang. "Interlamellar CA1 Network in the
Hippocampus." *Proceedings of the National Academy of Sciences of the United States of
America* 111, no. 35 (September 2014): 12919–24.
Young, Paul J., Anna B. Harper, Chris Huntingford, Nigel D. Paul, Olaf Morgenstern, Paul
A. Newman, Luke D. Oman, Sasha Madronich, and Rolando R. Garcia. "The Montreal
Protocol Protects the Terrestrial Carbon Sink." *Nature* 596, no. 7872 (2021): 384–88.
Zaidel, Dahlia W. "Art and Brain: The Relationship of Biology and Evolution to Art."
Progress in Brain Research 204 (2013): 217–33.
Zhou, Chunfang, and Lingling Luo. "Group Creativity in Learning Context: Understand-
ing in a Social-Cultural Framework and Methodology." *Creative Education* 3, no. 4
(2012): 392–99.
Zuidema, Willem, and Arie Verhagen. "What Are the Unique Design Features of
Language? Formal Tools for Comparative Claims." *Adaptive Behavior* 18, no. 1 (2010):
48–65.

INDEX

Page numbers in *italics* represent illustrations or tables.

dopamine, reward prediction error and, 46–49, *48*

Dostrovsky, Jonathan, 25

DYNA algorithms: for memory consolidation, 61; reinforcement learning and, 61–62; simulation-selection model and, 62; value learning and, 62

educational settings, creativity in, 141–42

EEG. *See* electroencephalography

Einstein, Albert, 126, 139; on power of imagination, 2

electroencephalography (EEG), 31

empirical abstract thinking, 77

encoding, of new memories, 4, 10

entorhinal cortex, 168–69; abstract thinking and, 77–78; grid cells in, 77–79; neocortex and, 77–79, *79*

episodic memory, 12; precuneus and, 104; retrieval of, 104

evolution: of dentate gyrus, 169; life diversity and, 71–72; of vertebrate land animals, *148*

evolution, of hippocampus, 51; in avian species, 63–64, *65, 66*; in bats, 69–70; in Chordata phylum, 71–72; cognitive maps and, 69; convergent, 71; imagination and, 63–72; of mammals, 63–64, *65, 66*; of neocortex, 85; place cells, 69–70; proximal explanations for, 63; simulation-selection model and, 67–68, 85–86; in tufted titmouse, 70–71; in Urchordata subphylum, 71–72; in Vertebrata subphylum, 71–72; in whales, 69–70

evolutionary lag hypothesis, 151–52

excitatory neurons, 178n2

executive function, 89; premotor cortex, 88; supplementary motor cortex, 88

expected return, 50

expensive tissue hypothesis, 85

explicit memory, 12

extinction, of species: climate change and, 150, *153*; great dying, 150; for *Homo sapiens*, 150; mass, 150–51; rates for, *150*

exuberant imagination, 139

face-selective neurons, 111–13, *113*

faculty of language in broad sense (FLB), 133–34

faculty of language in narrow sense (FLN), 133–34

fake visual images, false memory influenced by, 20–22, *21*

false memory: brain regions associated with, 16–17; coerced internalized false confessions as, 19; elaboration of information in, 20–21; fake visual images as influence on, 20–22, *21*; filled-in information, 17; Franklin, George, and, 17–18; Ingram case, 18–19; Loftus and, 19–20; "lost in a shopping mall" test and, 19–20; misinformation effect and, 19; Ofshe experiment, 19; partial, 21; repressed memory compared to, 18; satanic ritual and, 19; scope of, 16–17

Fitch, Tecumseh, 133

FLB. *See* faculty of language in broad sense

flexibility, prefrontal cortex and, 91–92, 94

FLN. *See* faculty of language in narrow sense

flow, creativity and, 143

fMRI. *See* functional magnetic resonance imaging

fossil fuels, use of: Carboniferous period and, 151–52, 193n6; climate change as result of, *151*, 151–52; evolutionary lag hypothesis and, 151–52

FOXP$_2$ gene, human language and, 131

Franklin, Eileen, 17–18

Franklin, George, 20; false memory and, 17–18

frontopolar cortex, 94, *95*; brain imaging for, *95*; concept cells and, 112

functional cell assembly, CA$_3$ neural circuit and, 40

functional magnetic resonance imaging (fMRI), 31; of hippocampus activation, 13; sharp-wave ripples and, 32

Gage, Phineas, 89–91, *90, 93*

Galilei, Galileo, 119

Geschwind, Norman, 125

imaging modalities:
electroencephalography, 31; for
frontopolar cortex, 95; functional
magnetic resonance imaging, 13, 31–32;
for human language, 128; magnetic
encephalography, 31
implicit memory, 12
inertia, 119
information processing, neuroscience
and, 33
information storage, in brain, 16–17
Ingram, Paul, false memory and, 18–19
inhibitory neurons, 178n2
innate abstract thinking, 77–79
innovation: through abstract thinking,
3–4; future perspectives on, 147, 149–61;
Homo sapiens and, 4; human capacity
for, 2, 147; imagination as promoter of,
2; optimistic perspectives on, 154–60;
pessimistic perspectives on, 149–52, 154;
technological growth and, 155–57
insect species, 1
intelligence explosion, 156. *See also*
artificial intelligence
intentionality. *See* shared intentionality
Intergovernmental Panel on Climate
Change (IPCC), 152
internal mentation, default mode network
and, 14, 68
interventional approach, to CA$_1$ neural
circuit, 53
introspection, human capacity for, 1
intuition, abstract thinking and, 77
IPCC. *See* Intergovernmental Panel on
Climate Change

Jung, Rex, 138
Jeong, Yeonseok, 53–54

Kant, Immanuel, 77
Kepler, Johannes, 119
Kesner, Raymond, 169
Knight, Chris, 134
knowledge growth, *157*, 157–58. *See also*
intelligence explosion
koan-contemplation zen, for creativity,
144–45
Kurzweil, Ray, 156, 165

language, human: abstract thinking and,
80, 86, 127–28, 137; acquisition of,
120–21; animal capacity for, 122–24;
aphasia and, 124–25, 128; areas of brain
for, 124–26; brain imaging studies
for, 128; Broca's area and, 124–25, *125*,
137; cognitive functions and, 132–33;
creativity and, 121–22; discreteness in,
121; displacement and, 121; evolution
of, 130–32; faculty of language in broad
sense, 133–34; faculty of language in
narrow sense, 133–34; fossilized records
of, 129; *FOXP$_2$* gene, 131; grammar
rules in, 121; for information exchange,
133; laryngeal descent theory and,
129–30; linguistic relativity hypothesis,
127; mental, 127; mirror neuron
hypothesis and, 135–37; Neanderthals
and, 131; as neuroenhancement,
127–28; neuroscientific studies on,
126; origins of, 129–35, 137; primary
purpose of, 127; primates and, 122–23;
processing of, 125–26; productivity
and, 121; purpose of, 133; Sapir-Whorf
hypothesis, 127; scaffolding role for,
128–29; shared intentionality and,
132–34; sign language and, 122–23;
social nature of, 134; thought and,
126–29; uniqueness of, 3–4, 121–22,
129; universality of, 120–21; Wernicke-
Geschwind model, *125*, 125–26;
Wernicke's area, 124–25, *125*; written,
86. *See also* communication
language deprivation syndrome, 128
language of thought hypothesis, 127
laryngeal descent theory, 129–30
Lashley, Karl, 10
learning: reinforcement, 46–47, 50, 52,
138; supervised, 110; unsupervised,
110–11
Leborgne, Louis Victor, 124
Lee, Albert, 28
Lee, Jong Won, 60
Lee, Sedol, 156
Lee, Sung-Hyun, 53
linguistic relativity hypothesis, 127
Loftus, Elizabeth, 19–20
Logothetis, Nikos, 32

9 780231 213363